James Young Simpson

Archaeological Essays

James Young Simpson

Archaeological Essays

ISBN/EAN: 9783743358423

Manufactured in Europe, USA, Canada, Australia, Japa

Cover: Foto ©berggeist007 / pixelio.de

Manufactured and distributed by brebook publishing software (www.brebook.com)

James Young Simpson

Archaeological Essays

ARCHÆOLOGICAL ESSAYS

BY THE LATE

SIR JAMES Y. SIMPSON, BART.

M.D., D.C.L.

ONE OF HER MAJESTY'S PHYSICIANS FOR SCOTLAND, AND PROFESSOR OF MEDICINE AND MIDWIFERY IN THE UNIVERSITY OF EDINBURGH

EDITED BY

JOHN STUART, LL.D.

SECRETARY OF THE SOCIETY OF ANTIQUARIES OF SCOTLAND

VOL. I.

EDINBURGH

EDMONSTON AND DOUGLAS

PUBLISHERS TO THE SOCIETY OF ANTIQUARIES

MDCCCLXXII

Printed by R. & R. Clark, *Edinburgh.*

THE EDITOR'S PREFACE.

THE late Sir James Simpson, in the midst of his anxious professional labours, was wont to seek for refreshment in the pursuit of subjects of a historical and archæological character, and to publish the results in the Transactions of different Societies and in scientific journals.

Some of these papers are now scarce, and difficult of access; and a desire having been expressed in various quarters for their appearance in a collected and permanent form, I was consulted on the subject by Sir Walter Simpson, who put into my hands copies of the various essays, with notes on some of them by his father, which seemed to indicate that he himself had contemplated their republication.

Having for a long time been acquainted with their merits, I did not hesitate to express a strong opinion in favour of their publication; and I accepted with pleasure the duty of editing them, which Sir Walter requested me to perform.

The papers in question were the fruit of inquiries

begun indeed as a relief from weightier cares; but as it was not in their author's nature to rest satisfied with desultory and superficial results in his treatment of any subject, so his archæological papers more resemble the exhaustive treatises of a leisurely student, than the occasional efforts of one overwhelmed in professional occupations.

In the present work will be found all the more important archæological papers of Sir James Simpson, collected from the various sources indicated in the Table of Contents.

The subjects to the antiquities of which Sir James first directed his attention were connected with his own profession; but, as time went on, his interest in historical pursuits deepened and expanded, and the questions discussed by him became more varied.

It has been thought best to arrange the papers of a general historical scope in the first volume, and those connected with professional antiquities in the second; but readers, who may wish to trace the order in which they were written by the author, will find their various dates in the Table.

The first paper, entitled "Archæology, its Past and its Future Work," was prepared as a lecture to the Society of Antiquaries of Scotland. This was done with a care and elaboration which are not always associated with such efforts; and, whether in indicating the object and end of the archæological student's pursuits,—sketching the past progress of the study,—and specifying the lines of research from which Scottish inductive archæology may be expected to derive additional data and facts,—nothing more thoroughly practical could be desired; while in his resumé of the difficulties and enigmas peculiar to Scottish antiquities, he may be said to have left none of them untouched, his passing allusions being, in many instances, suggestive of their solution.

The paper on "An old Stone-roofed Cell or Oratory in the Island of Inchcolm" affords an instance of the author's careful observation, and his fertility of illustration. The humble structure in question, which, at the time when it first attracted Sir James Simpson's notice, was used as a pig-stye, had few external features to suggest the necessity of farther inquiry; but after his eye had become accustomed to the architecture of the early monastic cells in Ireland, its real character flashed upon him, and he found that his conclusions coincided with the facts of the early history of the island.

These he gleaned from many sources, but in grouping them into a picture he enriched his narrative with various instructive notes; as on the "Mos Scotticum" of our early buildings; a comparison of the ruin with the Irish oratories; notices of other Island Retreats of Saints, and of the Saints themselves. In one of these he gives an instructive reference to a passage in the original Latin text of Boece about the round tower of Brechin, which had been overlooked by his translator Bellenden, and so was now quoted for the first time.

A copy of this paper on Inchcolm having been sent to his friend Dr. Petrie of Dublin, author of the well-known essay on the "Early Ecclesiastical Architecture and Round Towers of Ireland," it was returned after a time, enriched with many notes and illustrations. In now reprinting the paper these have been added, and are distinguished from the author's notes by having the letter P annexed to them. The subject of the Inchcolm oratory was one about which this great man felt much interest, and on which he could speak from the abundance of his knowledge and experience. The notes are therefore of special value, as furnishing the latest views of the author on mooted points of Celtic Ecclesiology, while they are conspicuous for the modesty and candour which were combined with Dr. Petrie's vast learning on the subject.

Thus, in his work on the Round Towers, Dr. Petrie assigned "about the year 1020" as the date of the round tower of Brechin, but in one of the notes he corrects himself, and explains the origin of his mistake:—"The recollection of the error which I made, by a carelessness not in such matters usual with me, in assigning this date 1020, instead of between the years 971 and 994, as I ought to have done, has long given me annoyance, and a lesson never to trust to memory in dates; for it was thus I fell into the mistake. I had the year 1020 on my mind, which is the year assigned by Pinkerton for the writing of the *Chron. Pictorum*, and, without stopping to remember or to refer, I took it for granted that it was the year of Kenneth's death, or rather of his gift."

In writing of the Early Churches or Oratories of Ireland, Dr. Petrie stated in his Essay—"they had a single doorway always placed in the centre of the west wall." In one of his notes, now printed, he thus qualifies the statement:—"I should perhaps have written *almost* always. The very few exceptions did not at the moment occur to me." Again, Sir James Simpson having quoted a passage from Dr. Petrie's work, in which the writer ascribes the old small stone-roofed church at Killaloe to the seventh century, Dr. Petrie, in his relative note, adds—"but now considers as of the tenth, or perhaps eleventh."

To the paper on "Leprosy and Leper Hospitals in Scotland and England" is now added a series of additional "Historical Notices," prepared by Dr. Joseph Robertson, with the accuracy and research for which, as is well known, my early friend was conspicuous.

The origin of the tract on "Medical Officers in the Roman Army" is explained in the following note, prefixed to the first edition:—"A few years ago my late colleague, Sir George Ballingall, asked me—'Was the Roman Army provided with Medical Officers?' He was interested in the subject as Professor of Military Surgery, and told me that he had made, quite unsuccessfully, inquiries on the matter in various quarters, and at various persons. I drew up for him a few remarks, which were privately printed and circulated among his class at the time. The present essay consists of an extension of these remarks."

The essay on the monument called "THE CATSTANE" suggested an explanation, which naturally elicited divergent criticisms. Some of these appear to have occasionally engaged Sir James Simpson's attention; and from some unfinished notes among his papers, it seems plain that he meant to notice them in an additional communication to the Society of Antiquaries.

In these notes, after recapitulating at the outset the facts adduced in his first paper, Sir James proceeds:—"These points of evidence, I ventured to conclude, '*tend at least to render it probable*' that the Catstane is a monument to Vetta, the grandfather of Hengist and Horsa. But I did not consider the question as a settled question. I began and ended my paper by discussing this early Saxon origin of the monument as problematical and probable, but not fixed. At the same time, I may perhaps take the liberty of remarking, that both in archæology and history we look upon some questions as sufficiently fixed and settled, regarding which we have less inferential and direct proof than we have respecting this solution of the enigma respecting the Catstane. The idea, however, that it was possible for a monument to a historic Saxon leader to be found in Scotland of a date antecedent to the advent of Hengist and Horsa to the shores of Kent, was a notion so repugnant to many minds, that, very naturally, various arguments have been adduced against it, while some high authorities have declared in favour of it. In this communication I propose to notice briefly some of the leading arguments that have been latterly brought forward both against and for the belief that the Catstane commemorates the ancestor of the Saxon conquerors of Kent.

"1. One anonymous writer has maintained, that if the Catstane was a monument to the grandfather of Hengist and Horsa, the inscription upon it should not have read 'In hoc tumulo jacet Vetta f(ilius) Victi,' but, on the contrary, 'Victus filius Vettæ.' In other words, he holds that the inscription reverses the order of paternity as given by Bede, Nennius, etc.[1] But all this is simply and altogether a mistake on the part of the writer. All the ancient genealogies describe Hengist and Horsa as the sons of Victgils, Victgils as the son of Vetta, and Vetta as the son of Victus. The Catstane inscriptions give Vetta and Victus in exactly the same order. When I pointed out to the writer the mistake into which he had, perhaps inadvertently, fallen, he turned round, and argued that in such names the vowels *e* and *i* were more trustworthy as permanent elements than the consonants *c* and *t*.[2] He argued, in other words, that Vecta as a proper name would not be found

[1] "The monument reverses the order of paternity of the two individuals, making Wecta the son of Witta, instead of Witta the son of Wecta, in which all the old genealogies agree."—*Athenæum*, July 5, 1862, p. 17.

[2] "The vowel is far more distinctive of the two names than the difference of *c* and *t*, letters which were continually interchanged."—*Ibid.* August 2, 1862, p. 149.

spelled with an *i*. If it were never so spelled with an *i*, that circumstance was no argument in favour of the strange error of criticism into which the writer had fallen; but the fact is, that in the famous chapter of Bede's history, in which the names Hengist and Horsa, and their genealogies, first occur, there is an instance given, showing that, contrary to the opinion of this writer, a proper name having, like *Vetta*, the letter *e* as a component, *may* change it to *i*. For Bede, in telling us that the men of Kent and of the Isle of Wight (Cantuarii et Victuarii) were sprung from the Jutes, spells the Isle of Wight (Vecta) with an *e*, and the inhabitants of it (Victuarii) with an *i*.

"The same writer states it as his opinion that the lettering in the Catstane inscription is not so old as I should wish to make it. 'It is,' says he, 'in our opinion, of later date even than Hengist himself, both in the formula of the inscription and in the character of the writing.' Perhaps the writer's opinion upon such a point is not worth alluding to, as it is maintained by no proof. But Edward Lhuyd—one of the very best judges in such questions in former days—stated the lettering to be of the fourth or fifth century, without having any hypothesis to support or subvert by this opinion. And the best palæographer of our own times—Professor Westwood—is quite of the same idea as to the mere age of the inscription, as drawn from its palæography and formula, an idea in which he is joined by an antiquary who has worked much with ancient lettering—viz. Professor Stephens of Copenhagen."

Although it is to be regretted that the contemplated remarks were not completed, it may be doubted if the question admitted of much further illustration; and, however unlikely the conclusion may be that the inscription on the Catstane, VETTA F[ILIUS] VICTI, is a contemporary commemoration of the grandfather of Hengist and Horsa, it may not be easy to suggest a solution of the question free from difficulties as puzzling. At all events the palæographic features of the inscription seem plainly to associate it with a class of rude post-Roman monuments, of which we have a good many examples in different parts of the kingdom; and it may be remarked that Mr. Skene, who has made this period of our history a special study, after investigating, with his usual acumen, the evidence which exists to show that the Frisians had formed settlements in Scotland at a period anterior to that usually assigned for the arrival of the Saxons in England, has established the fact of the early settlement on our northern coasts of a people called by the general name of Saxons, but in reality an offshoot from the Frisians, whose principal seat was on the shores of the Firth of Forth, and on the whole thinks it not impossible that the Catstane may be the tomb of their first leader Vitta, son of Vecta, the traditionary grandfather of Hengist and Horsa.[1]

Besides the papers now printed, Sir James Simpson contributed many shorter essays and reviews of books to magazines and newspapers. He also prepared a memo-

[1] *Proceedings of the Society of Antiquaries of Scotland*, vol. iv. p. 181.

randum, printed in the second volume of the "Sculptured Stones of Scotland," of a reading of the inscription on a sculptured cross at St. Vigeans in Forfarshire.[1] At the time of the final adjustment of this paper Sir James was an invalid, and confined to his bed, and I well remember the extreme, almost fastidious, care bestowed by him on the proof-sheet, in the course of my frequent visits to his bedroom.

It sometimes happened also that a subject originally treated in a paper by Sir James Simpson required a volume to exhaust it. Thus, in the spring of 1864, he read to a meeting of the Society of Antiquaries of Scotland a "Notice of the Sculpturing of Cups and Concentric Rings on Stones and Rocks in various parts of Scotland;" but materials afterwards so grew on his hands that his original Notice came to be expanded into a volume of nearly 200 pages, with 36 illustrative plates. His treatment of this curious subject furnishes a model for such investigations.[2]

Setting out with a description of the principal types of the sculptures, he investigates the chief deviations which occur. He next classifies the various monuments on which the sculptures have been observed, as standing-stones, cromlechs, stones in chambered tumuli, and stones in sepulchral cists. Another chapter describes their occur-

[1] *The Sculptured Stones of Scotland*, vol. ii. Notices of the Plates, p. 71.

[2] *Archaic Sculpturings of Cups, Circles, etc., upon Stones and Rocks in Scotland, England, and other Countries.* Edin. 1867.

rence on stones connected with archaic habitations, as weems, fortified buildings, in and near ancient towns and camps, and on isolated rocks and stones. After a description of analogous sculptures in other countries, there is a concluding chapter of general inferences founded on the facts accumulated in the previous part of the volume.

On the occasion of a rapid journey to Liverpool, Sir James Simpson visited a stone circle at Calder, near that city, and detected the true character of the sculptures on the stones, a very imperfect note of which I had recently brought under his notice. An account of this monument, which he prepared for the Historic Society of Lancashire and Cheshire, is printed in the Transactions of that body for 1865, and the following passages are quoted from it :—" Many suggestions, I may observe, have been offered in regard to the intent and import of such lapidary cup and ring cuttings as exist on the Calder Stones ; but none of the theories proposed solve, as it seems to me, the hieroglyphic mystery in which these sculpturings are still involved. They are old enigmatical 'handwritings on the wall,' which no modern reader has yet deciphered. In our present state of knowledge with regard to them, let us be content with merely collecting and recording the facts in regard to their appearances, relations, localities, etc. ; for all early theorising will, in all probability, end only in error. It is surely better frankly to own that we know not what these markings mean (and possibly may never know it), rather than wander off into that vague mystification and conjecture which in former days often brought discredit on the whole study of archæology.

"But in regard to their probable era let me add one suggestion. These cup and ring cuttings have now been traced along the whole length of the British Isles, from Dorsetshire to Orkney, and across their whole breadth from Yorkshire in England to Kerry in Ireland; and in many of the inland counties in the three kingdoms. They are evidently dictated by some common thought belonging to some common race of men. But how very long is it since a common race—or successive waves even of a common race — inhabited such distant districts as I have just named, and spread over Great Britain and Ireland, from the English Channel to the Pentland Firth, and from the shores of the German Ocean to those of the Atlantic?"

The special value of the inductive treatment of the subject adopted by Sir James Simpson is here conspicuous; and although no decided conclusion was come to on the age and meaning of the sculptures, or the people by whom they were made, yet a reader feels that the utmost has been made of existing materials; and that, while nothing has been left untouched which could throw light on the question, a broad and sure foundation has been laid on which all subsequent research must rest.

One of the Appendices to this volume contains an account of some ancient sculptures on the walls of certain caves in Fife. The essay originally appeared as a communication to the Royal Society of Edinburgh in January 1866, and was also soon afterwards printed separately —"Inscribed to James Drummond, Esq., R.S.A., as a small token of the Author's very sincere friendship and esteem."

The discovery of these cave sculptures affords an instance of the thoroughness which Sir James carried into all his investigations. While engaged in the preparation of his original paper for the Society of Antiquaries on the Sculpturing of Cups and Rings, he wished to ascertain all the localities and conditions of their occurrence. After describing the sculptured circles and cups which had been found on the stones of weems and "Picts' Houses," he referred to the caves on the coast of Fife, which he suggested might be considered as natural weems or habitations. These he had visited in the hope of discovering cup-markings; and in one near the village of Easter Wemyss he discovered faded appearances of some depressions or cups, with small single circles cut on the wall, adding to his description — "Probably a more minute and extensive search in these caves would discover many more such carvings."

This was written in 1864; and when the paper then prepared had been expanded into the volume of 1867, the passage just quoted was accompanied by the following note :—"I leave this sentence as it was written above two years ago. Shortly after that period, I revisited Wemyss, to inspect the other caves of the district, and make more minute observations than I could do in my first hurried visit, and discovered on the walls of some of them many carvings of animals, 'spectacle ornaments,' and other symbols exactly resembling in type and character the similar figures represented on the ancient so-called sculptured stones of Scotland, and, like them, probably about a thousand years old."[1]

In like manner, after Sir Gardner Wilkinson had detected a concentric circle of four rings sculptured on the pillar called "Long Meg," at the great stone circle of Salkeld, in Cumberland, Sir James Simpson paid a visit to the monument, when his scrutiny was rewarded by the discovery on this pillar of several additional groups of sculptures.[2]

In his lecture on Archæology, Sir James Simpson has indicated two lines of research, from which additional data and facts for the elucidation of past times might be expected—viz. researches beneath the surface of the earth, and researches among older works and manuscripts. By the former he meant the careful and systematised examinations in which the spade and pickaxe are so important, and have done such service in late years, and from which Sir James expected much more; and by the latter the exploring and turning to account the many stores of written records of early times yet untouched.

Being impressed with the value of the charters of our old religious houses for historical purposes, he, shortly

[1] *British Archaic Sculpturings*, p. 126. [2] *Idem*, p. 20.

before his death, had a transcript made of the Chartulary of the Monastery of Inchcolm, with a design to edit it as one of a series of volumes of monastic records for the Society of Antiquaries of Scotland.

But the services of Sir James Simpson to the cause of archæological research are not to be measured by his written contributions, remarkable as these are. Perhaps it may be said that his influence was most pregnant in kindling a love of research in others, by opening their eyes to see how much yet lay undiscovered, and how much each person could do by judicious effort in his own neighbourhood. With this view he on various occasions delivered lectures on special subjects of antiquity, and among his papers I found very full notes of lectures on Roman antiquities, one of which, on the "Romans in Britain," he delivered at Falkirk in the winter of 1862.

For many years the house of Sir James Simpson was the rendezvous of archæological students; and it was one of his great pleasures to bring together at his table men from different districts and countries, but united by the brotherhood of a common pursuit, for the discussion of facts and the exchange of thought.

The friends who were accustomed to these easy reunions will not soon forget the radiant geniality of the host, and his success in stimulating the discussions most likely to draw out the special stores of his guests. Others also, who were associated with Sir James in the visits to historical sites which he frequently planned, in the retrospect of the pleasant hours thus spent will feel how vain it is to hope for another leader with the attractions which were combined in him.

In the course of his numerous professional journeys he acquired a wonderfully accurate knowledge of the early remains of different districts; and so contagious was his enthusiasm for their elucidation, that both the professional brethren with whom he acted, and his patients, were speedily found among his correspondents and allies.

His presence at the meetings of Archæological Societies was ever regarded as a pleasure and benefit. Besides the stated meetings of the Society of Antiquaries of Scotland, which he attended with comparative frequency, and where he ever took a share in the discussions, he was present on various occasions at Congresses of the Archæological Institute, the Cambrian Association, and other kindred bodies, by means of which he was enabled to maintain an intercourse with contemporary fellow-labourers in the archæological field, and to attain that familiarity with different classes of antiquities which he turned to such account in the discussion and classification of the early remains of Scotland.

I must not speak of the wonderful combination of qualities which were conspicuous in Sir James Simpson, alongside of those which I have mentioned. This may safely be left to the more competent hand of Professor Duns, from whose memoir of his early friend so much may be expected, and where a more general estimate of his character will naturally be found. Yet, in bringing together this series of Sir James Simpson's Archæological Essays, it seemed not unsuitable for me to express something of my admiration of the earnest truth-seeking spirit with which they were undertaken, as well as of the genius and research with which they were executed.

JOHN STUART.

CONTENTS OF VOLUME I.

ERRATA.

Vol. I. P. 93. *For* "Clanmacnoise," *read* "Clonmacnoise."

P. 138. *For* "muniuscule," *read* "minuscule."

P. 166. *For* "Advenenerunt," *read* "Advenerant."

ARCHÆOLOGY:

ITS PAST AND ITS FUTURE WORK.[1]

It has become a practice of late years in this Society for one of the Vice-Presidents to read an Annual Address on some topic or topics connected with Archæology. I appear here to-night more in compliance with this custom than with any hope of being able to state aught to you that is likely to prove either of adequate interest or of adequate importance for such an occasion.

In making this admission, I am fully aware that the deficiency lies in myself, and not in my subject. For truly there are few studies which offer so many tempting fields of observation and comment as Archæology. Indeed, the aim and the groundwork of the studies of the antiquary form a sufficient guarantee for the interest with which these studies are invested. For the leading object and intent of all his pursuits is—MAN, and man's ways and works, his habits and thoughts, from the earliest dates at which we can find his traces and tracks upon the earth, onward and forwards along the journey of past time. During this long journey, man has everywhere left scattered behind and around him innumerable relics, forming so many permanent impressions and evidences of his march and progress. These impressions and evidences the

[1] An inaugural Address delivered to the Society of Antiquaries of Scotland, Session 1860-61.

antiquary searches for and studies—in the changes which have in successive eras taken place (as proved by their existing and discoverable remains) in the materials and forms of the implements and tools which man has from the earliest times used in the chase and in agriculture; in the weapons which he has employed in battle; in the habitations which he has dwelt in during peace, and in the earth-works and stone-works which he has raised during war; in the dresses and ornaments which he has worn; in the varying forms of religious faith which he has held, and the deities that he has worshipped; in the sacred temples and fanes which he has reared; in the various modes in which he has disposed of the dead; in the laws and governments under which he has lived; in the arts which he has cultivated; in the sculptures which he has carved; in the coins and medals which he has struck; in the inscriptions which he has cut; in the records which he has written; and in the character and type of the languages in which he has spoken. All the markings and relics of man, in the dim and distant past, which industry and science can possibly extract from these and from other analogous sources, Archæology carefully collects, arranges, and generalises, stimulated by the fond hope that through such means she will yet gradually recover more and more of the earlier chronicles and lost annals of the human race, and of the various individual communities and families of that race.

The objects of antiquarian research embrace events and periods, many of which are placed within the era of written evidence; but many more are of a date long anterior to the epoch when man made that greatest of human discoveries—the discovery, namely, of the power of permanently recording words, thoughts, and acts, in symbolical and alphabetic writing. To some minds it has seemed almost chimerical for the archæologist to expect to regain to any extent a knowledge of the conditions and circumstances of man, and of the different nations of men, before human cunning had learned to collect and inscribe them on stone or brass, or had fashioned them into written or traditional records capable of being safely floated down the stream of time. But the modern history of Archæology, as well as the analogies of other allied pursuits, are totally against any such hopeless views.

Almost within the lifetime of some who are still amongst us, there has sprung up and been cultivated—and cultivated most successfully too—a science which has no written documents or legible inscriptions to guide it on its path, and whose researches are far more ancient in their object than the researches of Archæology. Its subject is an antiquity greatly older than human antiquity. It deals with the state of the earth and of the inhabitants of the earth in times immeasurably beyond the earliest times studied by the antiquary. In the course of its investigations it has recovered many strange stories and marvellous chronicles of the world and of its living occupants—long, long ages before human antiquity even began. But if Geology has thus successfully restored to us long and important chapters in the pre-Adamite annals of the world's history, need Archæology despair of yet deciphering and reading—infinitely more clearly than it has yet done—that far later episode in the drama of the past which opens with the appearance of man as a denizen of earth. The modes of investigating these two allied and almost continuous sciences—Geology and Archæology—are the same in principle, however much the two sciences themselves may differ in detail. And if Geology, in its efforts to regain the records of the past state of animal and vegetable life upon the surface of the earth, has attractions which bind the votaries of it to its ardent study, surely Archæology has equal, if not stronger claims to urge in its own behoof and favour. To the human mind the study of those relics by which the archæologist tries to recover and reconstruct the history of the past races and nations of man, should naturally form as engrossing a topic as the study of those relics by which the geologist tries to regain the history of the past races and families of the *fauna* and *flora* of the ancient world. Surely, as a mere matter of scientific pursuit, the ancient or fossil states of man should—for man himself—have attractions as great, at least, as the ancient or fossil states of plants and animals ; and the old Celt, or Pict, or Saxon, be as interesting a study as the old Lepidodendron or Ichthyosaurus.

Formerly, the pursuit of Archæology was not unfrequently regarded as a kind of romantic dilettanteism, as a collecting together of meaningless antique relics and oddities, as a greedy hoarding and storing up of rubbish and frivolities that were fit only for an old curiosity shop, and that were valued merely because they were old ;—while the essays and writings of the antiquary were looked down upon as disquisitions upon very profitless conjectures, and very solemn trivialities. Perhaps the objects and method in which antiquarian studies were formerly pursued afforded only too much ground for such accusations. But all this is now, in a great measure, entirely changed. Archæology, as tempered and directed by the philosophic spirit, and quickened with the life and energy of the nineteenth century, is a very different pursuit from the Archæology of our forefathers, and has as little relation to their antiquarianism as modern Chemistry and modern Astronomy have to their former prototypes—Alchemy and Astrology. In proof of this, I may confidently appeal to the good work which Archæology has done, and the great advances which it has struck out in different directions within the last fifty years. Within this brief period it has made discoveries, perhaps in themselves of as momentous and marvellous a character as those of which any other modern science can boast. Let me cite two or three instances in illustration of this remark.

Dating, then, from the commencement of the present century, Archæology has—amidst its other work—rediscovered, through the interpretation of the Rosetta-stone, the long-lost hieroglyphic language of Egypt, and has thus found a key by which it has begun—but only as yet begun—to unlock the rich treasure-stores of ancient knowledge which have for ages lain concealed among the monuments and records scattered along the valley of the Nile. It has copied, by the aid of the telescope, the trilingual arrow-headed inscriptions written 300 feet high upon the face of the rocks of Behistun; and though the alphabets and the languages in which these long inscriptions were "graven with a pen of iron and lead upon the rocks for ever," had been long dead and unknown, yet, by a kind of philological divination, Archæology has exorcised and resuscitated both; and from these dumb stones, and from the analogous inscriptions of Van, Elwend, Persepolis, etc., it has evoked official gazettes and royal contemporaneous annals of the deeds and dominions of Darius, Xerxes, and other Persian kings.

By a similar almost talismanic power and process, it has forced the engraved cylinders, bricks, and obelisks of the old cities of Chaldea and Babylonia—as those of Wurka, Niffer, Muqueyer, etc.—to repeat over again to this present generation of men the names of the ancient founders of their public buildings, and the wars and exploits of their ancient monarchs. It has searched among the shapeless mounds on the banks of the Tigris, and after removing the shroud of earth and rubbish under which "Nineveh the Great" had there lain entombed for ages, it has brought back once more to light the riches of the architecture and sculptures of the palaces of that renowned city, and shown the advanced knowledge of Assyria —some thirty long centuries ago—in mechanics and engineering, in working and inlaying with metals, in the construction of the optical lens, in the manufactory of pottery and glass, and in most other matters of material civilisation. It has lately, by these and other discoveries in the East, confirmed in many interesting points, and confuted in none, the truth of the Biblical records. It has found, for instance, every city in Palestine and the neighbouring kingdoms whose special and precise doom was pronounced by the sure word of Prophecy, showing the exact state foretold of them twenty or thirty centuries ago,—as Askelon tenantless, the site of ancient Gaza "bald," old Tyre "scraped" up, and Samaria with its foundations exposed, and its "stones poured down in heaps" into the valley below. It has further, within the last few years, stolen into the deserts of the Hauran, through the old vigilant guard formed around that region by the Bedouin Arabs, and there—(as if in startling contradiction to the dead and buried cities of Syria, etc.)—it has—as was equally predicted—discovered the numerous cyclopic cities of Bashan standing perfect and entire, yet " desolate and without any to dwell therein,"—cities wrapped, as it were, in a state of mortal trance, and patiently awaiting the prophesied period of their future revival and rehabitation ; some of them of great size, as Um-el-Jemâl (probably the Beth-gamul of Scripture), a city covering as large a space as Jerusalem, with its high and massive basaltic town walls, its squares, its public buildings, its paved streets, and its houses with their rooms, stairs, revolving and frequently sculptured stone-doors, all nearly as complete and unbroken, as if its old inhabitants had only deserted it yesterday. Again, from another and more distant part of the East,—from the plains of India,—Archæology has recently brought to Europe, and at an English press printed for the first time, upwards of 1000 of the sacred hymns of the Rig-Veda, the most ancient literary work of the Aryan or Indo-European race of mankind ; for, according to the calm judgment of our ripest Sanskrit scholars, these hymns were composed before Homer sung of the wrath of Achilles ; and they are further remarkable, on this account, that they seem to have been transmitted down for upwards of 3000 years by oral tradition alone—the Brahmin priests up to the present day still spending—as Cæsar tells us the old Druidical priests of Gaul spent —twelve, twenty, or more years of their lives, in learning by heart these sacred lays and themes, and then teaching them in turn to their pupils and successors.

The notices of antiquarian progress in modern times, that I have hitherto alluded to, refer to other continents than our own. But since the commencement of the present century Archæology has been equally active in Europe. It has, by its recent devoted study of the whole works of art belonging to Greece, shown that in many respects a livelier and more familiar knowledge of the ancient inhabitants of that classic land is to be derived from the contemplation of their remaining statues, sculptures, gems, medals, coins, etc., than by any amount of mere school-grinding at Greek words and Greek quantities. It has recovered at the same time some interesting objects connected with ancient Grecian history; having, for example, during the occupation of Constantinople in 1854 by the armies of England and France, laid bare to its base and carefully copied the inscription, engraved some twenty-three centuries ago, upon the brazen stand of the famous tripod which was dedicated by the confederate Greeks to Apollo at Delphi, after the defeat of the Persian host at Platea,—an inscription that Herodotus himself speaks of, and by which, indeed, the Father of History seems to have authenticated his own battle-roll of the Greek combatants. Archæology has busied itself also, particularly of late years, in disinterring the ruins of numerous old Roman villas, towns, and cities in Italy, in France, in Britain, and in the other western colonies of Rome; and by this measure it has gained for us a clearer and nearer insight into every-day Roman life and habits, than all the wealth of classic literature supplies us with. Though perfectly acquainted with the Etruscan alphabet, it has hitherto utterly failed to read a single line of the numerous inscriptions found in Etruria, but yet among the unwritten records and relics of the towns and tombs of that ancient kingdom, it has recovered a wonderfully complete knowledge of the manners, and habits, and faith, of a great and prosperous nation, which—located in the central districts of Italy—was already far advanced in civilisation and refinement long before that epoch when Romulus is fabled to have drawn around the Palatine the first boundary line of the infant city which was destined to become the mistress of the world. Latterly, among all the western and northern countries of Europe, in Germany, in Scandinavia, in Denmark, in France, and in the British Islands, Archæology has made many careful and valuable collections of the numerous and diversified implements, weapons, etc., of the aboriginal inhabitants of these parts, and traced by them the stratifications, as it were, of progress and civilisation, by which our primæval ancestors successively passed upwards through the varying eras and stages of advancement, from their first struggles in the battle of life with tools of stone, and flint, and bone alone, till they discovered and applied the use of metals in the arts alike of peace and war; from those distant ages in which, dressed in the skins of animals, they wore ornaments made of sea-shells and jet, till the times when they learned to plait and weave dresses of hair, wool, and other fibres, and adorned their chiefs with torcs and armlets of bronze, silver, and gold. Archæology also has sought out and studied the strongholds and forts, the land and lake habitations of these, our primæval Celtic and Teutonic forefathers:—and has discovered among their ruins many interesting specimens of the implements they used, the dresses that they wore, the houses they inhabited, and the very food they fed upon. It has descended also into their sepulchres and tombs, and there — among the mysterious contents of their graves and cinerary urns—it has found revealed many other wondrous proofs of their habits and condition during this life, as well as of their creeds and faith in regard to a future state of existence.

By the aid of that new and most powerful ally, Comparative Philology, Archæology has lately made other great advances. By proofs exactly of the same linguistic kind as those by which the modern Spanish, French, and other Latin dialects can be shown to have all radiated from Rome as their centre, the old traditions of the eastern origin of all the chief nations of Europe have been proved to be fundamentally true ; for by evidence so "irrefragable" (to use the expression of the Taylorian professor of modern languages at Oxford), that "not an English jury could now-a-days reject it," Philological Archæology has shown that of the three great families of mankind—the Semitic, the Turanian, and the Aryan—this last, the Aryan, Japhetic, or Indo-European race, had its chief home about the centre of Western Asia ;—that betimes there issued thence from its paternal hearths, and wended their way southward, human swarms that formed the nations of Persia and Hindustan ;—that at distant and different, and in some cases earlier periods, there hived off from the same parental stock other waves of population, which wandered westward, and formed successively the European nations of the Celts, the Teutons, the Italians, the Greeks, and the Sclaves ;—and that while each exodus of this western emigration, which followed in the wake of its fellow, drove its earliest predecessor before it in a general direction further and further towards the setting sun, at the same time some aboriginal, and probably Turanian races, which previously inhabited portions of Europe, were gradually pushed and pressed aside and upwards, by the more powerful and encroaching Aryans, into districts either so sterile or so mountainous and strong, that it was too worthless or too difficult to follow them further—their remnants being represented at the present day by the Laps, the Basques, and the Esths. Philological Archæology has further demonstrated that the vast populations which now stretch from the mouth of the Ganges to the Pentland Firth,—sprung, as they are, with a few exceptions only, from the same primitive Aryan stock,—all use words which, though phonetically changed, are radically identical for many matters, as for the nearest relationships of family life, for the naming of domestic animals, and other common objects. Some of these archaic words indicate, by their hoary antiquity, the original pastoral employment and character of those that formed the parental stock in our old original Asiatic home; the special term, for example (the "pasu" of the old Sanskrit or Zend), which signified "private" property among the Aryans, and which we now use under the English modifications, "peculiar" and "pecuniary"—primarily meaning "flocks;"[1] the Sanskrit word for Protector, and ultimately for the king himself, "go-pa," being the old word for cowherd, and consecutively for chief herdsman; while the endearing name of "daughter" (the duhitar of the Sanskrit, the θυγάτηρ of the Greek), as applied in the leading Indo-European languages to the female children of our households, is derived from a verb which shows the original signification of the appellation to have been the "milker" of the cows. At the same time the most ancient mythologies and superstitions, and appa-

[1] As an illustration of this primitive pastoral idea of wealth, Dr. Livingstone told me, that on more than one occasion, when Africans were discoursing with him on the riches of his own country and his own chiefs at home, he was asked the searching and rather puzzling question, "But how many cows has the Queen of England?"

rently even the legends and traditions of the various and diversified Indo-European races, appear also, the more they are examined, to betray more and more of a common parentage. Briefly, and in truth, then, Philological Archæology proves that the Saxon and the Persian, the Scandinavian and the Greek, the Icelander and the Italian, the fair-skinned Scottish Highlander, and his late foe, the swarthy Bengalee, are all distant, very distant, cousins, whose ancestors were brothers that parted company with each other long, long ages ago, on the plains of Iran. That the ancestors of these different races originally lived together on these Asiatic plains "within the same fences, and separate from the ancestors of the Semitic and Turanian races," is (to quote the words of Max Müller), "a fact as firmly established as that the Normans of William the Conqueror were the Northmen of Scandinavia."

Lastly, to close this too long, and yet too rapid and imperfect sketch of some of the work performed by modern inductive Archæology, let me merely here add,—for the matter is too important to omit,—that, principally since the commencement of this century, Archæology has sedulously sat down among the old and forbidding stores of musty, and often nearly illegible manuscripts, charters, cartularies, records, letters, and other written documents, that have been accumulating for hundreds of years in the public and private collections of Europe, and has most patiently and laboriously culled from them annals and facts having the most direct and momentous bearing upon the acts and thoughts of our mediæval forefathers, and upon the events and persons of these mediæval times. By means of this last type of work, the researches of the antiquary have to a wonderful degree both purified and extended the history of this and of the other kingdoms of Europe. These researches have further, and in an especial manner, thrown a new flood of light upon the inner and domestic life of our ancestors, and particularly upon the conditions of the middle and lower grades of society in former times, — objects ever of primary moment to the researches of Archæology in its services as the workman and the pioneer of history. For, truly, human history, as it has been hitherto usually composed, has been too often written as if human chronicles ought to detail only the deeds of camps and courts—as if the number of men murdered on particular battle-fields, and the intrigues and treasons perpetrated in royal and lordly antechambers, were the sum total of actual knowledge which it was of any moment to transmit from one generation of men to another. In gathering, however, from the records of the past his materials for the true philosophy of history, the archæologist finds—and is now teaching the public to find—as great an attraction in studying the arts of peace as in studying the arts of war; for in his eyes the life, and thoughts, and faith of the merchant, and craftsman, and churl, are as important as those of the knight, and nobleman, and prince—with him the peasant is as grand and as genuine a piece of antiquity as the king.

Small in extent, scant in population, and spare in purse, as Scotland confessedly is, yet, in the cultivation of Archæology she has in these modern times by no means lagged behind the other and greater kingdoms of Europe. This observation is attested by the rich and valuable Museum of Scottish antiquities which this Society has gathered together—a Museum which, exclusively of its large collection of foreign coins, now numbers above 7000 specimens, for nearly 1000 of which we stand indebted to the enlightened zeal and patriotic munificence of one Scottish gentleman, Mr. A. Henry Rhind of Sibster. The same fact is attested also by the highly valuable character of the systematic works on Scottish Archæology which have been published of late years by some of our colleagues, such as the masterly *Pre-historic Annals of Scotland*, by Professor Daniel Wilson ; the admirable volume on *Scotland in the Middle Ages*, by Professor Cosmo Innes ; and the delightful *Domestic Annals of Scotland*, by Mr. Robert Chambers. The essays also, and monographs on individual subjects in Scottish Archæology, published by Mr. Laing, Lord Neaves, Mr. Skene, Mr. Stuart, Mr. Robertson, Mr. Fraser, Captain Thomas, Mr. Burton, Mr. Napier, Mr. M'Kinlay, Mr. M'Lauchlan, Dr. Wise, Dr. J. A. Smith, Mr. Drummond, etc., all strongly prove the solid and successful interest which the subject of Scottish Archæology has in recent times created in this city. The recent excellent town and county histories published by Dr. Peter Chalmers, Messrs. Irving, Jeffrey, Jervise, Pratt, Black, Miller, etc., afford evidences to the same effect. Nor can I forget in such an enumeration the two complete *Statistical Accounts of Scotland*. But if I were asked to name any one circumstance, as proving more than another the attention lately awakened among our countrymen by antiquarian inquiries, I would point, with true patriotic pride, to the numerous olden manuscript chronicles of Scotland, of Scottish towns, and Scottish monasteries, institutions, families, and persons, which have been printed within the last forty years—almost all of them having been presented as free and spontaneous contributions to Scottish Archæology and History by the members of the Bannatyne, the Abbotsford, the Maitland,

and the Spalding Clubs; and the whole now forming a goodly series of works extending to not less than three hundred printed quarto volumes.

But let us not cheat and cozen ourselves into idleness and apathy by reflecting and rejoicing over what has been done. For, after all, the truth is, that Scottish Archæology is still so much in its infancy, that it is only now beginning to guess its powers, and feel its deficiencies. It has still no end of lessons to learn, and perhaps some to unlearn, before it can manage to extract the true metal of knowledge from the ore and dross of exaggeration in which many of its inquiries have become enveloped. At this present hour we virtually know far less of the Archæology and history of Scotland ten or fifteen centuries ago than we know of the Archæology and history of Etruria, Egypt, or Assyria, twenty-five or thirty centuries ago.

In order to obtain the light which is required to clear away the dark and heavy mists which thus obscure the early Archæology of Scotland, how should we proceed? In the pursuits and investigations of Archæology, as of other departments of science, there has never yet been, and never will be discovered, any direct railway or royal road to the knowledge which we are anxious to gain, but which we are inevitably doomed to wait for and to work for. The different branches of science are Gordian knots, the threads of which we can only hope to unwind and evolve by cautious assiduity, and slow, patient industry. Their secrets cannot be summarily cut open and exposed by the sword of any son of Philip. But, in our day-dreams, it is not unpleasant sometimes to imagine the possibility of such a feat. It was, as we all know, very generally believed, in distant antiquarian times, that occasionally dead men could be induced to rise, and impart all sorts of otherwise unattainable information to the living. This creed, however, has not been limited to those ancient times, for, in our own days, many sane persons still profess to believe in the possibility of summoning the spirits of the departed from the other world back to this sublunary sphere. When they do so, they have always hitherto, as far as I have heard, encouraged these spirits to perform such silly juggling tricks, or requested them to answer such trivial and frivolous questions, as would seem to my humble apprehension to be almost insulting to the grim dignity and solemn character of any respectable and intelligent ghost. If, like Owen Glendower, or Mr. Home, I had the power to "call spirits from the vasty deep," and if the spirits answered the call, I—being a practical man—would fain make a practical use of their presence. Methinks I should feel grossly tempted, for example, to ask such of them as had the necessary foreknowledge, to rap out for me, in the first instance, the exact state of the English funds, or of the London stock and share-list, a week or a month hence; for such early information would, I opine —if the spirits were true spirits—be rather an expeditious and easy mode of filling my coffers, or the coffers of any man who had the good sense of plying these spiritual intelligences with one or two simple and useful questions. If, however, the spirits refused to answer such golden interrogatories as involving matters too mercenary and not sufficiently ghostly in their character, then I certainly should next ask them—and I would of course select very ancient spirits for the purpose—hosts of questions regarding the state of society, religion, the arts, etc., at the time when they themselves were living denizens of this earth. Suppose, for a moment, that our Secretaries, on summoning the next meeting of this Society, had the power of announcing in their billets that, by "some feat of magic mystery," a very select and intelligent deputation of ancient Britons and Caledonians, Picts, Celts, and Scots, and perhaps of Scottish Turanians, were to be present in our Museum—(certainly the most appropriate room in the kingdom for such a reunion)—for a short sederunt, somewhere between twilight and cock-crowing, to answer any questions which the Fellows might choose to ply them with, what an excitement would such an announcement create! How eagerly would some of our Fellows look forward to the results of one or two such "Hours with the Mystics." And what a battery of quick questions would be levelled at the members of this deputation on all the endless problems involved in Scottish Archæology. I think we may readily, and yet pretty certainly, conjecture a few of the questions, on our earlier antiquities alone, that would be put by various members that I might name, as:—

What is the signification of the so-called "crescent" and "spectacle" ornaments, and of the other unique symbols that are so common upon the 150 and odd ancient Sculptured Stones scattered over the north-eastern districts of Scotland?

What is the true reading of the old enigmatic inscriptions upon the Newton and St. Vigeau's stones, and of the Oghams on the stones of Logie, Bressay, Golspie, etc.?

Had Solinus Polyhistor, in the fourth century, any ground for stating that an ancient Ulyssean altar, written with Greek letters, existed in the recesses of Caledonia?

Who were Vetta, Victus, Memor, Loinedinus, Liberalis, Florentius, Mavorius, etc., whose names are recorded on the Romano-British monuments at Kirkliston, Yarrow, Kirkmadrine, etc., and what is the date of these monuments?

By what people was constructed the Devil's Dyke, which runs above fifty miles in length from Loch Ryan into Nithsdale?

When, and for what purpose, was the Catrail dug?

Was it on the line of the Catrail, or of the Roman wall between the Forth and Clyde, or on what other ground, that there was fought the great battle or siege of Cattracth or Kaltraez, which Aneurin sings of in his *Gododin*, and where, among the ranks of the British combatants, were "three hundred and sixty-three chieftains wearing the golden torcs" (some specimens of which might yet perhaps be dug up on the battle-field by our Museum Committee, seeing three only of these chiefs escaped alive); and how was the "bewitching mead" brewed, that Aneurin tells us was far too freely partaken of by his British countrymen before and during this fierce struggle with the Saxon foe?

Is the poet Aneurin the same person as our earliest native prose historian Gildas, the two appellations being relatively the Cymric and Saxon names of the same individual? Or were they not two of the sons or descendants of Caw of Cwm Cawlwyd, that North British chief whose miraculous interview with St. Cadoc near Bannawc (Stirlingshire?) is described in the life of that Welsh saint?

Of what family and rank was the poet—Merddin Wyllt—or "Merlin the Wild," who, wearing the chieftain's golden torc, fought at the battle of Arderydd, about A.D. 573, against Rhydderch Hael, that king of Alcluith or Dumbarton, who was the friend of St.

Columba, and "the champion of the (Christian) faith," as Merlin himself styles him? And when that victory was apparently the direct means of establishing this Christian king upon the throne of Strathclyde, and the indirect means which led to the recall of St. Kentigern from St. Asaph's to Glasgow, how is it that the Welsh Triads talk of it enigmatically as a battle for a lark's nest?

If Ossian is not a myth, when and where did he live and sing? Was he not an Irish Gael? And could any member of the deputation give us any accurate information about our old nursery friend Fingal or Fin Mac Coul? Was he really, after all, not greater, or larger, or any other than simply a successful and reforming general in the army of King Cormac of Tara, and the son-in-law of that monarch of Ireland?

From what part of Pictland did King Cormac obtain, in the third century, the skilled mill-wright, Mac Lamha, to build for him that first water-mill which he erected in Ireland, on one of the streams of Tara? And is it true, as some genealogists in this earthly world believe, that the lineal descendants of this Scottish or Pictish mill-wright are still millers on the reputed site of this original Irish water-mill?

The apostate Picts (*Picti apostati*) who along with the Scots are spoken of by St. Patrick in his famous letter against Coroticus, as having bought for slaves some of the Christian converts kidnapped and carried off by that chief from Ireland, were they inhabitants of Galloway, or of our more northern districts? And was the Irish sea not very frequently a "middle passage" in these early days, across which St. Patrick himself and many others were carried from their native homes and sold into slavery?

Was it a Pictish or Scottish, a British or a Roman architect that built "Julius' howff," at Stenhouse (*Stone-house*) on the Carron, and what was its use and object?

Were our numerous "weems," or underground houses, really used as human abodes, and were they actually so very dark, that when one of the inmates ventured on a joke, he was obliged—as suggested by "Elia"—to handle his neighbour's cheek to feel if there was any resulting smile playing upon it?

When, and by whom were reared the Titanic stone-works on the White Caterthun, and the formidable stone and earth forts and walls on the Brown Caterthun, on Dunsinane, on Barra, on the Barmekyn of Echt, on Dunnichen, on Dunpender, and on the tops of hundreds of other hills in Scotland?

How, and when, were our Vitrified Forts built? Was the vitrification of the walls accidental, or was it not rather intentional, as most of us now believe? In particular, who first constructed, and who last occupied the remarkable Vitrified Forts of Finhaven in Angus, and of the hill of Noath in Strathbogie? Was not the Vitrified Fort of Craig-Phadric, near Inverness, the residence of King Brude, the son of Meilochon, in the sixth century; and if so, is it true, as stated in the Irish Life of St. Columba, that its gates were provided with iron locks?

When, by whom, and for what object, were the moats of Urr, Hawick, Lincluden, Biggar, and our other great circular earth mounds of the same kind, constructed? Were they used for judicial and legal purposes, like the old Things of Scandinavia; and as the Tinwald Mount in the island of Man is used to this day? And were not some of them military or sepulchral works?

Who fashioned the terraces at Newlands in Tweeddale; and what was the origin of the many hillside terraces scattered over the country?

What is the age of the rock-caves of Ancrum, Hawthornden, etc., and were they primarily used as human habitations?

The sea-cave at Aldham on the Firth of Forth—when opened in 1831, with its paved floor strewed with charred wood, animal bones, limpet-shells, and apparently with a rock-altar at its mouth, having its top marked with fire, ashes adhering to its side, and two infants' skeletons lying at its base—was it a human habitation, or a Pagan temple?

What races sleep in the chambered barrows and cairns of Clava, Yarrows, Broigar, and in the many other similar old Scottish cities and houses of the dead?

By whom and for what purpose or purposes were the megalithic circles at Stennis, Callernish, Leys, Achnaclach, Crichie, Kennethmont, Midmar, Dyce, Kirkmichael, Deer, Kirkbean, Lochrutton, Torhouse, etc., etc., reared?

What were the leading peculiarities in the religious creed, faith, and festivals of Broichan and the other Caledonian or Pictish Magi before the introduction of Christianity?

When Coifi, the pagan high-priest of Edwin, the king of Northumbria and the Lothians, was converted to Christianity by Paulinus, in A.D. 627, he destroyed, according to Bede, the heathen idols, and set fire to the heathen temples and altars; but what was the structure of the pagan temples here in these days, that he could burn them,—while at the same time they were so uninclosed, that men on horseback could ride into them, as Coifi himself did after he had thrown in the desecrating spear?

Was not our city named after this Northumbrian Bretwalda, "Edwin's-burgh?" Or was the Eiddyn of which Aneurin speaks before the time of Edwin, and the Dinas Eiddyn that was one of the chief seats of Llewddyn Lueddog (Lew or Loth), the grandfather of St. Kentigern or Mungo of Glasgow, really our own Dun Edin? Or if the Welsh term "Dinas" does not necessarily imply the high or elevated position of the place, was it Caer Eden (Cariden, or Blackness), at the eastern end of the Roman Wall, on the banks of the Forth?

Did our venerable castle rock obtain the Welsh name of Din or Dun Monaidh, from its being "the fortress of the hill," and was its other Cymric appellation Agnedh, connected with its ever having been given as a marriage-portion (Agwedh)? Or did its old name of Maiden Castle, or Castrum Puellarum, not rather originate in its olden use as a female prison, or as a school, or a nunnery?

And is it true, as asserted by Conchubhranus, that the Irish lady Saint, Darerca or Monnine, founded, late in the fifth century, seven churches (or nunneries?) in Scotland, on the hills of Dun Edin, Dumbarton, Stirling, Dunpelder, and Dundevenal, at Lanfortin near Dundee, and at Chilnacasc in Galloway?

When, and by whom, were the Round Towers of Abernethy, Brechin, and Eglishay built? Were there not in Scotland or its islands other such *"turres rotundae mirâ arte constructae,"* to borrow the phrase of Hector Boece regarding the Brechin tower?

If St. Patrick was, as some of his earliest biographers aver, a Strathclyde Briton, born about A.D. 387 at Nempthur (Nemphlar, on the Clyde?) and his father Calphurnius was, as St. Patrick himself states in his Confession, a deacon, and his grandfather Potitus a priest, then he belonged to a family two generations of which were already office-bearers in Scotland in the Christian Church;—but were there many, or any such families in Scotland before St. Ninian built his stone church at Whithern about A.D. 397, or St. Palladius, the missionary of Pope Celestine, died about A.D. 431, in the Mearns? And was it a mere rhetorical flourish, or was there some foundation for the strong and distinct averment of the Latin father Tertullian, that, when he wrote, about the time of the invasion of Scotland by Severus (*circa* A.D. 210), there were places in Britain beyond the limits of the Roman sway already subject to Christ?

When Dion Cassius describes this invasion of Scotland by Severus, and the Roman Emperor's loss of 50,000 men in the campaign, does he not indulge in "travellers' tales," when he further avers that our Caledonian ancestors were such votaries of hydropathy that they could stand in their marshes immersed up to the neck in water for live-long days, and had a kind of prepared homœopathic food, the eating of a piece of which, the size of a bean, entirely prevented all hunger and thirst?

Cæsar tells us that dying the skin blue with woad was a practice common among our British ancestors some 1900 years ago;—are Claudian and Herodian equally correct in describing the very name of Picts as being derived from a system of painting or tattooing the skin, that was in their time as fashionable among some of our Scottish forefathers, as it is in our time in New Zealand, and among the Polynesians?

According to Cæsar, the Britons wore a moustache on the upper lip, but shaved the rest of the beard; and the sole stone—fortunately a fragment of ancient sculpture—which has been saved from the ruins of the old capital of the Picts at Forteviot, shows a similar practice among them. But what did they shave with? Were their razors of bronze, or iron, or steel? And where, and by whom, were they manufactured?

Was the state of civilisation and of the arts among the Caledonians, when Agricola invaded them, about A.D. 80 or 81, as backward as some authorities have imagined, seeing that they were already so skilled in, for example, the metallurgic arts, as to be able to construct, for the purposes of war,—chariots, and consequently chariot-wheels, long swords, darts, targets, etc.?

As the swords of the Caledonians in the first century were, according to Tacitus, long, large, and blunt at the point, and hence in all probability made of iron, whence came the sharp-pointed leaf-shaped bronze swords so often found in Scotland, and what is the place and date of their manufacture? Were they earlier? And what is the real origin of the large accumulation of spears and other instruments of bronze, some whole, and others twisted, as if half-melted with heat, which, with human bones, deer and elk-horns, were dredged up from Duddingston Loch about eighty years ago, and constituted, it may be said, the foundation of our Museum? Was there an ancient bronze-smith shop in the neighbourhood; or were these not rather the relics of a burned crannoge that had formerly existed in this lake, within two miles of the future metropolis of Scotland?

Could the deputation inform us where we might find, buried and concealed in our muirs or mosses, and obtain for our Museum some interesting antiquarian objects which we sadly covet—such as a specimen or two, for instance, of those Caledonian spears described by Dion, that had a brazen apple, sounding when struck, attached to their lower extremity? or one of those statues of Mercury that, Cæsar says, were common among the Western Druids? or one of the *covini* mentioned by Tacitus—(for we are anxious to know if its wheels were of iron or bronze; how these wheels made, as Cæsar tells us the wheels of the British war-chariots made, a loud noise in running; and whether or not they had, as some authorities maintain, scythes or long swords affixed to their axles)? or where we might dig up another specimen of such ancient and engraved silver armour as was some years ago discovered at Norrie's Law, in Fife, and unfortunately melted down by the jeweller at Cupar? or could any of the deputation refer us to any spot where we might have a good chance of finding a concealed example of such glass goblets as were, according to Adamnan, to be met with in the royal palace of Brude, king of the Picts, when St. Columba visited him, in A.D. 563, in his royal fort and hall (*munitio, aula regalis*) on the banks of the Ness?

Whence came King "Cruithne," with his seven sons, and the Picts? Were they of Gothic descent and tongue, as Mr. Jonathan Oldbuck maintained in rather a notorious dispute in the parlour at Monkbarns? or were they "genuine Celtic," as Sir Arthur Wardour argued so stoutly on the same memorable occasion?

Were the first Irish or Dalriadic Gaeidhil or Scots who took possession of Argyll (*i.e.*, Airer-Gaeidheal, or the district of the Gaeidhel), and who subsequently gave the name of Scot-land to the whole kingdom, the band of emigrants that crossed from Antrim about A.D. 506 under the leadership of Fergus and the other sons of Erc; or, as the name of "Scoti" recurs more than once in the old sparse notices of the tribes of the kingdom before this date, had not an antecedent colony, under Cairbre Riada, as stated by Bede, already passed over and settled in Cantyre a century or two before?

Our Reformed British Parliament is still so archæological as to listen, many times each session, to Her Majesty, or Her Majesty's Commissioners, assenting to their bills, by pronouncing a sentence of old and obsolete Norman French—a memorial in its way of the Norman Conquest; and our State customs are so archæological that, when Her Majesty, and a long line of her illustrious predecessors, have been crowned in Westminster Abbey, the old Scottish coronation-stone, carried off in A.D. 1296 by Edward I. from Scone, and which had been previously used for centuries as the coronation-stone of the Scotic, and perhaps of the Irish, or even the Milesian race of kings, has been placed under their coronation-chair—playing still its own archaic part in this gorgeous state drama. But is this Scone or Westminster coronation-stone really and truly—as it is reputed to be by some Scottish historians—the famous *Lia Fail* of the kings of Ireland, that various old Irish writings describe as formerly standing on the Hill of Tara, near the Mound of the Hostages? Or does not the *Lia Fail*—"the stone that roared under the feet of each king that took possession of the throne of Ireland"—remain still on Tara—(though latterly degraded to the office of a grave-stone)—as is suggested by the distinguished author of the History and Antiquities of Tara Hill? If any of our deputies from ghostdom formerly belonged to the court of Fergus MacErc, or originally sailed across with him in his fleet of *currachs*, perhaps they will be so good as tell us if in reality the royal or any other of the accompanying skin-canoes was ballasted then or subsequently with a sacred stone from Ireland, for the coronation of our first Dalriadic king; and especially would we wish it explained to us how such a precious monument as the *Lia Fail* of Tara was or could be smuggled away by such a small tribe as the Dalriadic Scots at first were? Perhaps it would be right and civil to tell the deputation at once, that the truth is we are anxious to decide the knotty question as to whether the opinions of Edward I. or of Dr. Petrie are the more correct in regard to this "Stone of Fate?" Or if King Edward was right politically, is Dr. Petrie right archæologically, in his views on this subject? In short, does the *Lia Fail* stand at the present day—as is generally believed—in the vicinity of the Royal Halls of Westminster, or in the vicinity of the Royal Halls of Tara?

What ancient people, destitute apparently of metal tools and of any knowledge of mortar, built the gigantic burgs or duns of Mousa, Hoxay, Glenelg, Carloway, Bragar, Kildonan, Farr, Rogart, Olrick, etc., with galleries and chambers in the thickness of their huge uncemented walls? Is it true, as the Irish bardic writers allege, that some of the race of the Firbolgs escaped, after the battle at one of the Moyturas to the Western Islands and shores of Scotland, and that thence, after several centuries, they were expelled again by the Picts, after the commencement of the Christian era, and subsequently returned to the coast of Galway, and built, or rebuilt, there and then, the great analogous burgs of Dun Ængus, Dun Conchobhair, etc., in the Irish isles of Aran?[1]

[1] As some confirmation of the views suggested in the preceding question,

What is the signification of those mysterious circles formed of diminishing concentric rings which are found engraved, sometimes on rocks outside an old aboriginal village or camp, as at Rowtin Lynn and Old Bewick; sometimes on the walls of underground chambers, as in the Holm of Papa Westray, and in the island of Eday; sometimes on the walls of a chambered tumulus, as at Pickaquoy in Orkney; or on the interior of the lid of a kistvaen, as at Craigie Hall, near Edinburgh, and probably also at Coilsfield and Auchinlary; or on a so-called Druidical stone, as on "Long Meg" at Penrith?

Is it true that a long past era—and, if so, at what era—our predecessors in this old Caledonia had nothing but tools and implements of stone, bone, and wood? Are there no gravel-beds in Scotland in which we could probably find large deposits of the celts and other stone weapons—with bored and worked deer-horns, of that distant stone-age—such as have been discovered on the banks of the Somme and the Loire in France? And were the people of that period in Scotland Celtic or pre-Celtic?

When the first wave of Celtic emigrants arrived in Scotland, did they not find a Turanian or Hamitic race already inhabiting it, and were those Scottish streams, lakes, etc., which bear, or have borne, in their composition, the Euskarian word *Ura* (water) —as the rivers Urr, Orr, and Ury, lochs Ur, Urr, and Orr, Urrquhart, Cath-Ures, Or-well, Or-rea, etc., named by these Turanian aborigines?

We know that in Iona, ten or twelve centuries ago, Greek was

my friend Captain Thomas pointed out to me, after the Address was given, that the name of the fort in St. Kilda was, as stated by Martin and Macaulay, "Dun Fir-bholg."

written, though we do not know if the Iona library possessed—what Queen Mary had among the sixteen Greek volumes[1] in her library—a copy of Herodotus; but we are particularly anxious to ascertain if the story told by Herodotus of Rhampsinitus, and the robbery of his royal treasury by that "Shifty Lad" "the Master Thief,"[2] was in vogue as a popular tale among the Scottish Gaels or Britons in the oldest times? The tale is prevalent in different guises from India to Scotland and Scandinavia among the Aryans, or alleged descendants of Japhet; Herodotus heard it about twenty-three centuries ago in Egypt, and consequently (according, at least, to some high philologists), among the alleged descendants of Shem; and could any Scottish Turanians, as alleged descendants of Ham, in the deputation, tell us whether the tale was also a favourite with them and their forefathers? For if so, then, in consonance with the usual reasoning on this and other popular tales, the story must have been known in the Ark itself, as the sons of Noah separated soon after leaving it, and yet all their descendants were acquainted with this legend. But have these and other such simple tales not originated in many different places, and among many different people, at different times; and have they not an appearance of similarity, merely because, in the course of their development, the earliest products of the human fancy, as well as of the human

[1] Including the works of Homer, Plato, Sophocles, etc. Her library catalogue shows also a goodly list of "Latyn Buikis," and classics. In a letter to Cecil, dated St. Andrews, 7th April 1562, Randolph incidentally states that Queen Mary then read daily after dinner "somewhat of Livy" with George Buchanan.

[2] See these stories in Mr. Dasent's *Norse Tales*, and in Mr. Campbell's collection of the *Popular Tales of the West Highlands*.

hand, are always more or less similar under similar circumstances?

Or perhaps, passing from more direct interrogatories, we might request some of the deputation to leave with us a retranslation of that famous letter preserved by Bede, which Abbot Ceolfrid addressed about A.D. 715 to Nectan III., King of the Picts, and which the venerable monk of Jarrow tells us was, immediately after its receipt by the Pictish King and court, carefully interpreted into their own language? or to be so good as write down a specimen of the Celtic or Pictish songs that happened to be most popular some twelve or fourteen centuries ago? or describe to us the limits at different times of the kingdoms of the Strathclyde Britons and Northumbrians, and of the Picts and Dalriadic Scots? or fill up the sad gaps in Mr. Innes' map of Scotland in the tenth century, containing, as it does, the names of one river only, and some thirteen Scottish church establishments and towns; or tell us where the "urbs Giudi" and the Pictish "Niduari" of Bede were placed, and why Ængus the Culdee speaks (about A.D. 800) of Cuilenross, or Culross, as placed in Strath-h-Irenn in the Comgalls, between Slieve-n-Ochil and the Sea of Giudan? or identify for us the true sites of the numerous rivers, tribes, divisions, and towns—or merely perhaps stockaded or rathed villages—which Ptolemy in the second century enters in his geographical description of North Britain? or particularise the precise bounds of the Meatæ and Attacotti, and of the two Pictish nations mentioned by Ammianus Marcellinus, namely, the Dicaledonæ and Vecturiones? or trace out for us the course of Agricola's campaigns in Scotland, especially marking the exact site of the great victory of the Mons Grampius, and thus deciding at once and for ever whether the two enormous cairns placed above the moor of Ardoch cover the remains of the 10,000 slain ; or whether the battle was fought at Dealgin Ross, or at Findochs, or at Inverpeffery, or at Urie Hill in the Mearns, or at Mormond in Buchan, or at the "Kaim of Kinprunes?" which last locality, however, was, it must be confessed, rather summarily and decisively put out of Court some time ago by the strong personal evidence of Edie Ochiltree.

If these, and some thousand-and-one similar questions regarding the habits, arts, government, language, etc., of our Primæval and Mediæval Forefathers could be at once summarily and satisfactorily answered by any power of "gramarye," then the present and the future Fellows of the Society of Antiquaries of Scotland would be saved an incalculable amount of difficult investigation and hard work. But unfortunately I, for one at least, have no belief that any human power can either unsphere the spirits of the dead for a night's drawing-room amusement, or seduce the "wraiths" of our ancestors to "revisit the glimpses of the moon" even for such a loyal and patriotic object as the furtherance of Scottish Archæology. Nevertheless I doubt not, at the same time, that many of these supposed questions on the dark points of Scottish antiquities will yet betimes be answered more or less satisfactorily. But the answers, if ever obtained, will be obtained by no kind of magic except the magic of accumulated observations, and strict stern facts ;—by no necromancy except the necromancy of the cautious combination, comparison, and generalisation of these facts ;—by no enchantment, in short, except that special form of enchantment for the advancement of every science which the mighty and potent wizard—Francis Bacon—taught to his fellow-men, when he taught them the spell-like powers of the inductive philosophy.

The data and facts which Scottish antiquaries require to seek out and accumulate for the future furtherance of Scottish Archæology, lie in many a different direction, waiting and hiding for our search after them. On some few subjects the search has already been keen, and the success correspondingly great. Let me specify one or two instances in illustration of this remark.

As a memorable example, and as a perfect Baconian model for analogous investigations on other corresponding topics—in the way of the full and careful accumulation of all ascertainable premises and data before venturing to dogmatise upon them—let me point to the admirable work of Mr. Stuart on the Sculptured Stones of Scotland—an almost national work, which, according to Mr. Westwood (the highest living authority on such a subject), is "one of the most remarkable contributions to Archæology which has ever been published in this or any other country."

"Crannoges"—those curious lake-habitations, built on piles of wood, or stockaded islands,—that Herodotus describes in lake Prasias, five or six centuries before the Christian era, constituting dwellings there which were then impregnable to all the military resources of a Persian army,—that Hippocrates tells us were also the types of habitation employed in his day by the Phasians, who sailed to them in single-tree canoes,—that in the same form of houses erected upon tall wooden piles, are still used at the present day as a favourite description of dwelling in the creeks and rivers running into the Straits of Malacca, and on the coasts of Borneo and New Guinea, etc., and the ruins of which have been found in numerous lakes in Ireland, England, Switzerland, Germany, Denmark, etc.;—Crannoges, I say, have been searched for and found also in various lochs in our own country; and the many curious data ascertained with regard to them in Scotland will be given in the next volume of our Society's proceedings by Mr. Joseph Robertson, a gentleman whom we all delight to acknowledge as pre-eminently entitled to wield amongst us the pen of the teacher and master in this as in other departments of Scottish antiquities.

Most extensive architectural data, sketches, and measurements, regarding many of the remains of our oldest ecclesiastical buildings in Scotland (including some early Irish Churches, with stone roofs and Egyptian doors, that still stand nearly entire in the seclusion of our Western Islands), have been collected by the indomitable perseverance and industry of Mr. Muir; and when the work which that most able ecclesiologist has now in the press is published, a great step will doubtless be made in this neglected department of Scottish antiquities.

In addition, however, to the assiduous collection of all ascertainable facts regarding the existing remains of our sculptured stones, our crannoges, and our early ecclesiastical buildings, there are many other departments of Scottish antiquities urgently demanding, at the hands of the numerous zealous antiquaries scattered over the country, full descriptions and accurate drawings of such vestiges of them as are still left—as, for example :—

I. Our ancient Hill-forts of Stone and Earth.

II. Our old cyclopic Burgs and Duns.

III. Our primæval Towns, Villages, and Raths.

IV. Our Weems or Underground Houses.

V. Our Pagan sepulchral Barrows, Cairns, and Cromlechs.

VI. Our Megalithic Circles and Monoliths.

VII. Our early Inscribed Stones ; etc. etc.

Good and trustworthy accounts of individual specimens, or groups of specimens, of most of these classes of antiquities, have been already published in our Transactions and Proceedings, and elsewhere. But Scottish Archæology requires of its votaries as large and exhaustive a collection as possible, with accurate descriptions, and, when possible, with photographs or drawings—or mayhap with models (which we greatly lack for our Museum)—of all the discoverable forms of each class ; as of all the varieties of ancient hill-strongholds ; all the varieties of our underground weems, etc. The necessary collection of all ascertainable types, and instances of some of these classes of antiquities, will be, no doubt, a task of much labour and time, and will in most instances require the combined efforts of many and zealous workers. This Society will be ever thankful to any members who will contribute even one or two stones to the required heap. But all past experience has shown that it is useless, and generally even hurtful, to attempt to frame hypotheses upon one, or even upon a few specimens only. In Archæology, as in other sciences, we must have full and accurate premises before we can hope to make full and accurate deductions. It is needless and hopeless for us to expect clear, correct, and philosophic views of the character and of the date and age of such archæological objects as I have enumerated, except by following the triple process of (1) assiduously collecting together as many instances as possible of each class of our antiquities; (2) carefully comparing these instances with each other, so as to ascertain all their resemblances and differences; and (3) contrasting them with similar remains in other cognate countries, where—in some instances, perhaps—there may exist, what possibly is wanting with us, the light of written history to guide us in elucidating the special subjects that may happen to be engaging our investigations—ever remembering that our Scottish Archæology is but a small, a very small, segment of the general circle of the Archæology of Europe and of the World.

The same remarks, which I have just ventured to make, as to the proper mode of investigating the classes of our larger archæological subjects, hold equally true also of those other classes of antiquities of a lighter and more portable type, which we have collected in our museums; such, for instance, as the ancient domestic tools, instruments, personal ornaments, weapons, etc., of stone, flint, bone, bronze, iron, silver, and gold, which our ancestors used; the clay and bronze vessels which they employed in cooking and carrying their food; the handmills with which they ground their corn; the whorls and distaffs with which they span, and the stuff and garments spun by them, etc. etc. It is only by collecting, combining, and comparing all the individual instances of each antiquarian object of this kind—all ascertainable specimens, for example, of our Scottish stone celts and knives; all ascertainable specimens of our clay vessels; of our leaf-shaped swords; of our metallic armlets; of our grain rubbers and stone-querns, etc. etc.—and by tracing the history of similar objects in other allied countries, that we will read aright the tales which these relics—when once properly interrogated—are capable of telling us of the doings, the habits, and the thoughts of our distant predecessors.

It is on this same broad and great ground—of the indispensable necessity of a large and perfect collection of individual specimens of all kinds of antiquities for safe, sure, and successful deduction—that we plead for the accumulation of such objects in our own or in other public antiquarian collections. And in thus pleading with the Scottish public for the augmentation and enrichment of our Museum, by donations of all kinds, however slight and trivial they may seem to the donors, we plead for what is not any longer the property of this Society, but what is now the property of the nation. The Museum has been gifted over by the Society of Antiquaries to the Government—it now belongs, not to us, but to Scotland—and we unhesitatingly call upon every true-hearted Scotsman to contribute, whenever it is in his power, to the extension of this Museum, as the best record and collection of the ancient archæological and historical memorials of our native land. We call for such a central general ingathering and repository of Scottish antiquities for another reason. Single specimens and examples of archæological relics are, in the hands of a private individual, generally nought but mere matters of idle curiosity and wild conjecture; while all of them become of use, and sometimes of great moment, when placed in a public collection beside their fellows. Like stray single words or letters that have dropt from out the Book of Time, they themselves, individually, reveal nothing, but when placed alongside of other words and letters from the same book, they gradually form—under the fingers of the archæologist—into lines, and sentences, and paragraphs, which reveal secret and stirring legends of the workings of the human mind, and human hand, in ages of which, perchance, we have no other existing memorials.

In attempting to read the cypher of these legends aright, let us guard against one fault which was unfortunately too often committed in former days, and which is perhaps sometimes committed still. Let us not fall into the mistake of fancying that everything antiquarian, which we do not see at first sight the exact use of, must necessarily be something very mysterious. Old distaff-whorls, armlets, etc., have, in this illogical spirit, been sometimes described as Druidical amulets and talismen ; ornamented rings and bosses from the ancient rich Celtic horse-harness, discovered in sepulchral barrows, have been published as Druidical astronomical instruments ; and in the last century some columnar rock arrangement in Orkney was gravely adduced by Toland as a Druidical pavement. It is this craving after the mysterious, this reprehensible irrationalism, that has brought, indeed, the whole subject of Druidism into much modern contempt with many archæologists. No doubt Druidism is a most interesting and a most important subject for due and calm investigation, and the facts handed down to us in regard to it by Cæsar, Diodorus, Mela, Strabo, Pliny, and other classic and hagiological authors, are full of the gravest archæological bearings ; but no doubt also many antiquarian relics, both large and small, have been provokingly called Druidical, merely because their origin and object were unknown. We have not, for instance, a particle of direct evidence for the too common belief that our stone circles were temples which the Druids used for worship; or that our cromlechs were their sacrificial altars. In fact, formerly the equanimity of the old theoretical class of archæologists was disturbed by these leviathan notions about Druids and Druidesses as much as the marine zoology of the poor sailor was long disturbed by his leviathian notions about sea-serpents and mermaids.

In our archæological inquiries into the probable uses and import of all doubtful articles in our museums or elsewhere, let us proceed upon a plan of the very opposite kind. Let us, like the geologists, try always, when working with such problems, to understand the past by reasoning from the present. Let us study backwards from the known to the unknown. In this way we can easily come to understand, for example, how our ancestors made those single-tree canoes, which have been found so often in Scotland, by observing how the Red Indian, partly by fire and partly by the hatchet, makes his analogous canoe at the present day; how our flint arrows were manufactured, when we see the process by which the present Esquimaux manufactures his; how our predecessors fixed and used their stone knives and hatchets, when we see how the Polynesian fixes and uses his stone knives and hatchets now; how, in short, matters sped in respect to household economy, dress, work, and war, in this old Caledonia of ours, during even the so-called Stone Age, when we reflect upon and study the modes in which matters are conducted in that new Caledonia in the Pacific—the inhabitants of which knew nothing of metals till they came in contact with Europeans, not many years ago; how, in long past days, hand and home-made clay vessels were the chief or only vessels used for cooking and all culinary purposes, seeing that in one or two parts of the Hebrides this is actually the state of matters still.

The collection of home-made pottery on the table—glazed with milk—is the latest contribution to our Museum. It was recently brought up, by Captain Thomas and Dr. Mitchell, from the parish of Barvas, in the Lewis. These "craggans," jars, or bowls, and other culinary dishes, are certainly specimens of the ceramic art in its most primitive state ;—they are as rude as the rudest of our old cinerary urns ; and yet they constitute, in the places in which they were made and used, the principal cooking, dyeing, and household vessels possessed by some of our fellow-countrymen in this the nineteenth century.[1] In the adjoining parish of Uig, Captain Thomas found and described to us, two years ago, in one of his instructive and practical papers, the small bee-hive stone houses in which some of the nomadic inhabitants of the district still live in summer. Numerous antiquarian remains, and ruins of similar houses and collections of houses, exist in Ireland, Wales, Cornwall, Switzerland, and perhaps in other kingdoms; but apparently they have everywhere been long ago deserted as human habitations, except in isolated and outlying spots among the Western Islands of Scotland. The study of human habits in these Hebridean houses, at the present day, enables us to guess what the analogous

[1] Among the people of the district of Barvas, most of them small farmers or crofters, a metal vessel or pot was a thing almost unknown twelve or fourteen years ago. Their houses have neither windows nor chimneys, neither tables nor chairs ; and the cattle and poultry live under the same roof with their human possessors. If a Chinaman or Japanese landed at Barvas, and went no further, what a picture might he paint, on his return home, of the state of civilisation in the British Islands.

human habits probably were, when, for example, the old Irish city of Fahan—consisting of similar structures only—was the busy scene of human life and activity in times long past. These, and other similar facts, besides teaching us the true road to some forms of archæological discovery, teach us also one other important lesson,—namely, that there are in reality two kinds of antiquity, both of which claim and challenge our attention. One of these kinds of antiquity consists in the study of the habits and works of our distant predecessors and forefathers, who lived on this earth, and perhaps in this segment of it, many ages ago. The other kind of antiquity consists of the study of those archaic human habits and works which may, in some corners of the world, be found still prevailing among our fellow-men—or even among our own fellow-countrymen—down to the present hour, in despite of all the blessings of human advancement, and the progress of human knowledge. By one kind of antiquity we trace the slow march and revolutions of centuries; by the other we trace the still slower march and revolutions of civilisation, in countries and kingdoms where the glittering theories of the politician might have led us to expect a different and a happier state of matters.

Besides the antiquarian relics of a visible and tangible form to which I have adverted, as demanding investigation and collection on our part, there are various antiquarian relics of a non-material type in Scottish Archæology which this Society might perhaps do much to collect and preserve, through the agency of active committees, and the assistance of many of our countrymen, who, I doubt not, could be easily incited to assist us in the required work. One of these matters is a fuller collection and digest than we yet possess of the old superstitious beliefs and practices of our forefathers. And certainly some strange superstitions do remain, or at least lately did remain, among us. The sacrifice, for example, of the cock and other animals for recovery from epilepsy and convulsions, is by no means extinct in some Highland districts. In old Pagan and Mithraic times we know that the sacrifice of the ox was common. I have myself often listened to the account given by one near and dear to me, who was in early life personally engaged in the offering up and burying of a poor live cow as a sacrifice to the Spirit of the Murrain. This occurred within twenty miles of the metropolis of Scotland. In the same district a relative of mine bought a farm not very many years ago. Among his first acts, after taking possession, was the inclosing a small triangular corner of one of the fields within a stone wall. The corner cut off—and which still remains cut off—was the "Goodman's Croft"—an offering to the Spirit of Evil, in order that he might abstain from ever blighting or damaging the rest of the farm. The clergyman of the parish, in lately telling me the circumstance, added, that my kinsman had been, he feared, far from acting honestly with Lucifer, after all, as the corner which he had cut off for the "Goodman's" share was perhaps the most worthless and sterile spot on the whole property. Some may look upon such superstitions and superstitious practices as matters utterly vulgar and valueless in themselves; but in the eyes of the archæologist they become interesting and important when we remember that the popular superstitions of Scotland, as of other countries, are for the most part true antiquarian vestiges of the pagan creeds and customs of our earlier ancestors; our present Folk-lore being merely in general a degenerated and debased form of the highest mythological and medical lore of very distant times. A collection of the popular superstitions and practices of the different districts of Scotland now, ere (like fairy and goblin forms vanishing before the break of day) they melt and disappear totally before the light and the pride of modern knowledge, would yet perhaps afford important materials for regaining much lost antiquarian knowledge. For as the palæontologist can sometimes reconstruct in full the types of extinct animals from a few preserved fragments of bones, possibly some future archæological Cuvier may one day be able to reconstruct from these mythological fragments, and from other sources, far more distinct figures and forms than we at present possess of the heathen faith and rites of our forefathers.

Perhaps a more important matter still would be the collection, from every district and parish of Scotland, of local lists of the oldest names of the hills, rivers, rocks, farms, and other places and objects; and this all the more that in this age of alteration and change many of these names are already rapidly passing away. Yet the possession of a Scottish antiquarian gazetteer or map of this kind would not only enable us to identify many localities mentioned in our older deeds and charters, but more—the very language to which these names belong would, perhaps, as philological ethnology advances, betimes serve as guides to lead our successors, if they do not lead us, to obtain clearer views than we now have of the people that aboriginally inhabited the different districts of our country, and the changes which occurred from time to time in these districts in the races which successively had possession of them. In this, as in other parts of the world, our mountains and other natural objects often obstinately retain, in despite of all subsequent changes and conquests, the appellations with which they were originally baptised by the aboriginal possessors of the soil ; as, for example, in three or four of the rivers which enter the Forth nearest to us here—viz., the Avon, the Amond, and the Esk on this side ; and the Dour, at Aberdour, on the opposite side of the Firth. For these are all old Aryan names, to be found as river appellations in many other spots of the world, and in some of its oldest dialects. The Amond or Avon is a simple modification of the present word of the Cymric "Afon," for "river," and we have all from our school-days known it under its Latin form of "Amnis." The Esk, in its various modifications of Exe, Axe, Uisk, etc., is the present Welsh word, "Uisk," for "water," and possibly the earliest form "asqua," of the Latin noun "aqua." Again, the noun "Dour"—Douro—so common an appellative for rivers in many parts of Europe, is, according to some of our best etymologists, identical with, or of the same Aryan source as the "Uda," or water, of the sanskrit, "ὕδωρ" of the Greeks, and the "Dwr" or "Dour" of the Cambrian and Gael. The archæologist, like the Red Indian when tracking his foe, teaches himself to observe and catch up every possible visible trace of the trail of archaic man ; but, like the Red Indian also, he now and again lays his ear on the ground to listen for any sounds indicating the presence and doings of him who is the object of his pursuit. The old words which he hears whispered in the ancient names of natural objects and places supply the antiquary with this kind of audible archæological evidence. For, when cross-questioned at the present day as to their nomenclature, many, I repeat, of our rivers and lakes, of our hills and headlands, do, in their mere names, telegraph back to us, along mighty distances of time, significant specimens of the tongue spoken by the first inhabitants of their district—in this respect resembling the doting and dying octogenarian that has left in early life the home of his fathers, to sojourn in the land of the stranger, and who remembers and babbles at last—ere the silver cord of memory is utterly and finally loosed—one language only, and that some few words merely, in the long unspoken tongue which he first learned to lisp in his earliest infancy.

The special sources and lines of research from which Scottish inductive Archæology may be expected to derive the additional data and facts which it requires for its elucidation are many and various. Let me here briefly allude to two only, and these two of rather opposite characters,—viz. (1), researches beneath the surface of the earth; and (2), researches among olden works and manuscripts.

In times past Scottish Archæology has already gained much from digging; and in times to come it is doubtless destined to gain yet infinitely more from a systematised use of this mode of research. For the truth is, that beneath the surface of the earth on which we tread—often not above two or three feet below that surface, sometimes not deeper than the roots of our plants and trees—there undoubtedly lie, in innumerable spots and places,—buried, and waiting only for disinterment,—antiquarian relics of the most valuable and important character. The richest and rarest treasures contained in some of our antiquarian museums have been exhumed by digging; and that digging has been frequently of the most accidental and superficial kind—like the discovery of the silver mines of Potosi through the chance uprooting of a shrub by the hand of a climbing traveller.

The magnificent twisted torc, containing some £50 worth of pure gold, which was exhibited in Edinburgh in 1856, in the Museum of the Archæological Institute, was found in 1848 in Needwood Forest, lying on the top of some fresh mould which had been turned up by a fox, in excavating for himself a new earth-hole. Formerly, on the sites of the old British villages in Wiltshire, the moles, as Sir Richard Hoare tells us, were constantly throwing up to the surface numerous coins and fragments of pottery. We are indebted to the digging propensities of another animal for the richest collection of silver ornaments which is contained in our Museum: For the great hoard of massive silver brooches, torcs, ingots, Cufic and other coins, etc., weighing some 16 lbs. in all, which was found in 1857 in the Bay of Skaill in Orkney, was discovered in consequence of several small pieces of the deposit having been accidentally uncovered by the burrowings of the busy rabbit. That hoard itself is interesting on this other account, that it is one of 130 or more similar silver deposits, almost all found by digging, that have latterly been discovered, stretching from Orkney, along the shores and islands of the Baltic, through Russia southward, towards the seat of the government of those Eastern Caliphs who issued the Cufic coins which generally form part of these collections —this long track being apparently the commercial route along which those merchants passed, who, from the seventh or eighth to the eleventh century, carried on the traffic which then subsisted between Asia and the north of Europe.

The spade and plough of the husbandman are constantly disinterring relics of high value to the antiquary and numismatist. The matchless collection of gold ornaments contained in the Museum of the Irish Academy has been almost entirely discovered in the course of common agricultural operations. The pickaxe of the ditcher, and of the canal and railway navvies, have often also, by their accidental strokes, uncovered rich antiquarian treasures. The remarkable massive silver chain, ninety-three ounces in weight, which we have in our Museum, was found about two feet below the surface, when the Caledonian Canal was dug in 1808. One of the largest gold armlets ever discovered in Scotland was disinterred at Slateford in cutting the Caledonian Railway. Our Museum contains only a model of it; for the original—like many similar relics, when they consisted of the precious metals—was sold for its mere weight in bullion, and lost—at least to Archæology—in the melting-pot of the jeweller, in consequence of the former unfortunate state of our law of treasure-trove. And it cannot perhaps be stated too often or too loudly, that such continued wanton destruction of these relics is now so far provided against; for by a Government ordinance, the finder of any relics in ancient coins, or in the precious metals, is now entitled by law, on delivering them up to the Crown for our National Museum, to claim "the full intrinsic value" of them from the Sheriff of the district in which they chance to be discovered—a most just and proper enactment, through the aid of which many such relics will no doubt be henceforth properly preserved.

But the results of digging to which I have referred are, as I have already said, the results merely of accidental digging. From a systematised application of the same means of discovery, in fit and proper localities, with or without previous ground-probing, Archæology is certainly entitled to expect most valuable consequences. The spade and pickaxe are become as indispensable aids in some forms of archæological, as the hammer is in some forms of geological research. The great antiquarian treasures garnered up in our sepulchral barrows and olden kistvaen cemeteries, are only to be recovered to antiquarian science by digging, and by digging, too, of the most careful and methodised kind. For in such excavations it is a matter of moment to note accurately every possible separate fact as to the position, state, etc., of all the objects exposed; as well as to search for, handle, and gather these objects most carefully. In excavating, some years ago, a large barrow in the Phœnix Park at Dublin, two entire skeletons were discovered within the chamber of the stone cromlech which formed the centre of the sepulchral mound. A flint knife, a flint arrow-head, and a small fibula of bone were found among the rubbish, along with some cinerary urns; but no bronze or other metallic implements. The human beings buried there had lived in the so-called Stone Period of the Danish archæologists. Some hard bodies were observed immediately below the head of one of the skeletons, and by very cautious and careful picking away of the surrounding earth, there was traced around the neck of each a complete necklace formed of the small sea-shells of the Nerita, with a perforation in each shell to admit of a string composed of vegetable fibres being passed through them. Without due vigilance how readily might these interesting relics have been overlooked!

The spade and mattock, however, have subserved, and will subserve, other important archæological purposes besides the opening of ancient cemeteries. They will probably enable us yet to solve to some extent the vexed question of the true character of our so-called "Druidical circles" and "Druidical stones," by proving to us that one of their uses at least was sepulchral. The bogs and mosses of Ireland, Denmark, and other countries, have, when dug into, yielded up great stores of interesting antiquarian objects —usually wonderfully preserved by the qualities of the soil in which they were immersed—as stone and metallic implements, portions of primæval costume, combs, and other articles of the toilet, pieces of domestic furniture, old and buried wooden houses, and even, as in the alleged case of Queen Gunhild, and other "bogged" or "pitted" criminals, human bodies astonishingly entire, and covered with the leathern and other dresses in which they died. All this forms a great mine of antiquarian research, in which little or nothing has yet been accomplished in Scotland. It is only by due excavations that we can hope to acquire a proper analytical knowledge of the primæval abodes of our ancestors,— whether these abodes were in underground "weems," or in those hitherto neglected and yet most interesting objects of Scottish Archæology, namely, our archaic villages and towns, the vestiges and marks of which lie scattered over our plains and mountain sides—always near a stream, or lake, or good spring—usually marked by groups of shallow pits or excavations (the foundations of their old circular houses) and a few nettles—generally protected and surrounded on one or more sides by a rath or earth-wall—often near a hill-fort—and having attached to them, at some distance in the neighbourhood, stone graves, and sometimes, as on the grounds about Morton Hall, monoliths and barrows.

Last year we had detailed at length to the Society the very remarkable results which Mr. Neish had obtained by simple persevering digging upon the hill of the Laws in Forfarshire, exposing, as his excavations have done, over the whole top of the hill, extensive Cyclopic walls of several feet in height, formerly buried beneath the soil, and of such strange and puzzling forms as to defy as yet any definite conjecture of their character. No doubt similar works, with similar remains of implements, ornaments, querns, charred corn, etc., will yet be found by similar diggings on other Scottish hills; and at length we may obtain adequate data for fixing their nature and object, and perhaps even their date. Certainly every Scotch antiquary must heartily wish that the excellent example of earnest and enlightened research set by Mr. Neish was followed by others of his brother landholders in Scotland.

At the present time the sites and remains of some Roman cities in England are being restored to light in this way—as the old city of Uriconium (Wroxeter), where already many curious discoveries have rewarded the quiet investigations that are being carried on;—and Borcovicus in Northumberland (a half-day's journey from Edinburgh), one of the stations placed along the magnificent old Roman wall which still exists in wonderful preservation in its neighbourhood, and itself a Roman town, left comparatively so entire that "Sandy Gordon" described it long ago as the most remarkable and magnificent Roman station in the whole island, while Dr. Stukely spoke of it enthusiastically as the "Tadmor of Britain." I was lately told by Mr. Longueville Jones, that in the vicinity of Caerleon—the ancient Isca Silurum of the Roman Itinerary—the slim sharpened iron rod used as a ground-probe had detected at different distances a row of buried Roman houses and villas, extending from the old city into the country for nearly three miles in length. Here, as elsewhere, a rich antiquarian mine waits for the diggings of the antiquary; and elsewhere, as here, the ground-probe will often point out the exact spots that should be dug, with far more certainty than the divining rod of any Dousterswivel ever pointed out hidden hoards of gold or hidden springs of water.

But it is necessary, as I have already hinted, to seek and hope for additional archæological materials in literary as well as in subterraneous researches. And certainly, one especial deficiency which we have to deplore in Scottish Archæology is the almost total want of written documents and annals of the primæval and early mediæval portions of Scottish history. The antiquaries of England and Ireland are much more fortunate in this respect than we are; for they possess a greater abundance of early documents than we can boast of. Indeed, after Tacitus' interesting account of the first Roman invasion of Scotland under Agricola, and a few meagre allusions to, and statements regarding this country and its inhabitants by some subsequent classic authors, we have, for a course of seven or eight centuries, almost no written records of any authority to refer to. The chief, if not the only, exceptions to this general remark, consist of a few scattered entries bearing upon Scotland in the Irish Annals—as in those of Tighernach and Ulster; some facts related by Bede; some statements given in the lives and legends of the early Scottish, Welsh, and Irish saints;[1] and various copies of the list of the Pictish kings.

[1] One of these Lives—that of St. Columba by Adamnan—has been annotated by Dr. Reeves with such amazing lore that it really looks as if the

When we come down beyond the eleventh and twelfth centuries, our written memorials rapidly increase in quantity and extent. I have already alluded to the fact that three hundred quarto volumes —nearly altogether drawn from unpublished manuscripts—have been printed by the Scottish clubs within the last forty years. Mr. Robertson informs me that in the General Register House alone (and independently of other and private collections), there is material for at least a hundred volumes more; and the English Record Office contains, as is well known, many unedited documents referring to the building of various Scottish castles by Edward I., and to other points interesting to Scottish Archæology and History. The Welsh antiquaries have obtained from the Government offices in London various important documents of this description referring to Wales. Why should the antiquaries of Scotland not imitate them in this respect?

Modern experience has shown that it is not by any means chimerical to expect, that we may yet recover, from various quarters, and from quite unexpected sources, too, writings and documents of much interest and importance in relation both to British and to Scottish Archæology. Of that great fossil city Pompeii, not one hundredth part, it is alleged, has as yet been fully searched; and, according to Sir Charles Lyell, the quarters hitherto cleared out are those where there was the least probability of discovering manuscripts. It would be almost hoping beyond the possibility of hope to expect that in some of its unexplored mansions, one of the rich libraries of those ancient Roman times may turn up, presenting

Editor had acquired his wondrous knowledge of ancient Iona and Scotland by some such "uncanny" aids as an archæological "deputation of spirits."

papyri deeply interesting to British antiquaries, and containing, for example, a transcript of that letter on the habits and character of the inhabitants of Britain which Cicero himself informs us that he desired his brother Quintus to write, when, as second in command, he accompanied Julius Cæsar in his first invasion of our island ;—or a copy of that account which Himilico the Carthaginian, had drawn up of his voyage, some centuries before the Christian era, to the Tin Islands, and other parts northwards of the Pillars of Hercules ;—or a roll of those Punic Annals which Festus Avienus tells us that he himself consulted when (probably in the fourth century) he wrote those lines in his "*Ora Maritima*" in which he gives a description of Great Britain and Ireland.

The antiquaries of Scotland would heartily rejoice over the discovery of lost documents far less ancient than these. Perhaps I could name two or three of our colleagues who would perfectly revel over the recovery, for instance, of one or two leaves of those old Pictish annals (*veteres Pictorum libri*) that still existed in the twelfth century, and in which, among other matters, was a brief account (once copied by the Pictish clerk Thana, the son of Dudabrach, for King Ferath, at Meigle) of the solemn ceremony which took place when King Hungus endowed the church of St. Andrews, in presence of twelve members of the Pictish regal race, with a grant of many miles of broad acres, and solemnly placed with his royal hands on the altar of the church a piece of fresh turf in symbolisation of his royal land-gift. We all deplore that we possess no longer what the Abbot Ailred of Rievaulx, and the monk Joceline of Furness possessed, namely, biographies, apparently written in the old language of our country, of two of our earliest

Scottish saints—St. Ninian of Whithorn, and St. Kentigern of Glasgow ; and we grieve that we have lost even that Life of St. Serf, which, along with a goodly list of service and other books (chained to the stalls and desks), was placed, before the time of the Reformation, in the choir of the Cathedral of Glasgow, as we know from the catalogue which has been preserved of its library.

But let us not at the same time forget that Scottish archæological documents, as ancient as any of these, have been latterly rediscovered, and rediscovered occasionally in the most accidental way ; and let us not, therefore, despair of further, and perhaps even of greater success in the same line. Certainly the greatest of recent events in Scottish Archæology was the casual finding, within the last two or three years, in one of the public libraries at Cambridge, of a manuscript of the Gospels, which had formerly belonged to the Abbey of Deer, in Aberdeenshire. The margin and blank vellum of this ancient volume contain, in the Celtic language, some grants and entries reaching much beyond the age of any of our other Scottish charters and chronicles. The oldest example of written Scottish Gaelic that was previously known was not earlier than the sixteenth century. Portions of the Deer Manuscript have been pronounced by competent scholars to be seven centuries older.

The most ancient known collection of the laws of Scotland—a manuscript written about 1270—was detected in the public library of Berne, and lately restored to this country. In 1824, Mr. Thomson, a schoolmaster at Ayr, picked up, on an old bookstall in that town, a valuable manuscript collection of Scotch burghal laws written upwards of four centuries ago.

Sometimes, as in this last instance, documents of great value in Scottish Archæology have made narrow escapes from utter loss and destruction.

I was told by the late Mr. Thomas Thomson—a gentleman to whom we are all indebted for promoting and systematising our studies—that a miscellaneous, but yet in some points valuable collection of old vellum manuscripts was left, at the beginning of the present century, by a poor peripatetic Scottish tailor, who could not read one word of the old black letter documents which he spent his life and his purse in collecting. Being a visionary claimant to one of the dormant Scottish peerages, he buoyed himself up with the bright hope that some clever lawyer would yet find undoubted proofs of his claims in some of the written parchments which he might procure. Sir Robert Cotton is said to have discovered one of the original vellum copies of the Magna Charta in the shop of another tailor, who, holding it in his hand, was preparing to cut up this charter of the liberties of England into tape for measuring some of England's sons for coats and trousers. The missing manuscript of the History of Scotland, from the Restoration to 1681, which was written by Sir George Mackenzie, the King's Advocate, was rescued from a mass of old paper that had been sold for shop purposes to a grocer in Edinburgh. Some fragments of the Privy Council Records of Scotland—now preserved in the General Register House—were bought among waste snuff-paper.[1] Occasionally even a very small preserved fragment of an

[1] This alludes to the portion of a mutilated volume for the year 1605, which came into Mr. Laing's hands, and was given by him to the Deputy Clerk Register. But singular enough, as Mr. Laing has since informed me,

ancient document has proved of importance. Mr. Robertson informs me that, in editing the old Canons of the Scottish Church, he has derived considerable service from a single leaf of a contemporary record of the Canons of the sixteenth century, which had been used and preserved in the old binding of a book. This single leaf is the only bit of manuscript of the Scotch sixteenth century Canons that is known to exist in Scotland.

In 1794 eight official volumes of the Scottish Secretary of State's Register of Seisins were discovered in a bookseller's shop in Edinburgh, after they had remained concealed for more than 185 years.

Among the great mass of interesting Scottish manuscripts preserved in our General Register House, there is one dated Arbroath, —April 1320 ;—perhaps the noblest Scottish document of that era. It is the official duplicate of a letter of remonstrance addressed to Pope John XXII. by the Barons, Freeholders, and Community of Scotland, in which these doughty Scotsmen declare, that so long as a hundred of them remain alive, they will never submit to the dominion of England. This venerable record and precious declaration of Scottish independence, written on a sheet of vellum, and authenticated by the dependant seals of its patriotic authors, was detected by a deceased Scottish nobleman in a most precarious

the identical MS. of Sir George Mackenzie, above noticed, was brought to him for sale as probably a curious volume ; it having by some accident been *a second time sold for waste paper !* Having no difficulty in recognising the volume, he of course secured it, and, agreeably to the expressed intention of the Editor of the work in 1821, the MS. has been deposited in the Advocates' Library, where, it is to be hoped, it may now remain in safety.

situation ; for he discovered it ruthlessly stuck into the fire-place of his charter-room.

Contested points in Scottish Archæology and history have been occasionally settled by manuscript discoveries that were perfectly accidental.

After the blowing up of the Kirk of the Field, the only one of Darnley's servants that escaped was brought by the Earl of Murray before the English Council, and there gave evidence, implying that Queen Mary—that ever-interesting princess, who has been doubtlessly both over-decried by her foes and over-praised by her friends—was cognisant of the intended murder of her husband, inasmuch as, beforehand, she ordered an old bed to be placed in Darnley's room, and the richer bed that previously stood in it to be removed. Nearly three hundred years after that dark and sordid insinuation was made, a roll of papers was casually found, during a search among some legal documents of the early part of the seventeenth century, and one of the leaves in that roll contained a contemporary and authenticated official return of the royal furniture lost by the blowing up of the King's residence. Among other items, this leaf proved, beyond the possibility of further cavil, that the bed which stood in Darnley's room was, up to the time of his death, unchanged, and was not, as alleged by Mary's enemies, an old and worthless piece of furniture, but, on the contrary, was "a bed of violet velvet, with double hangings, braided with gold and silver (ung lictz de veloux viollet a double pante passemente dor et argent)."

The finest old Teutonic cross in Scotland is the well-known pillar which stands in the churchyard of Ruthwell, in Dumfriesshire. It was ignominiously thrown down, by a decree of the General

Assembly of the Presbyterian Church, in 1642; but its broken fragments were collected, as far as possible, and the cross itself again erected, by the late clergyman of the parish, Dr. Henry Duncan, who published in the Transactions of this Society correct drawings of the Runic inscription on this ancient monument. Two Danish antiquaries, Repp and Finn Magnusen, tried to read these Runic lines, and tortured them into very opposite, and let me simply add, very ridiculous meanings, about a grant of land and cows in Ashlafardhal, and Offa, a kinsman of Woden, transferring property to Ashloff, etc., all which they duly published. That great antiquary and Saxon scholar, the late Mr. Kemble, then happened to turn his attention to the Ruthwell inscription, and saw the runes or language to be Anglo-Saxon, and in no ways Scandinavian, as had been supposed. He found that the inscription consisted of a poem, or extracts from a poem, in Anglo-Saxon, in which the stone cross, speaking in the first person, described itself as overwhelmed with sorrow because it had borne Christ raised upon it at His crucifixion, had been stained with the blood poured from His side, and had witnessed His agonies,—

> " I raised the powerful King,
> The Lord of the heavens;
> I dared not fall down," etc. etc.

Who was to decide between the very diverse opinions, and still more diverse readings, of this inscription by the English antiquary and his Danish rivals? An accidental discovery in an old manuscript may be justly considered as having settled the whole question. For, two or three years after Mr. Kemble had published his reading of the inscription, the identical Anglo-Saxon poem which he had found written on the Ruthwell cross was casually discovered in an extended form under the title of the "Dream of the Holy Rood." The old MS. volume of Saxon homilies and religious lays from which the book containing it was printed, was found by Dr. Blum in a library at Vercelli, in Italy.

With these rambling remarks I have already detained you far too long. Ere concluding, however, bear with me for a minute or two longer, while I shortly speak of one clamant subject—viz. the strong necessity of this Society, and of every Scotsman, battling and trying to prevent, if possible, the further demolition of the antiquarian relics scattered over Scotland.

Various human agencies have been long busy in the destruction and obliteration of our antiquarian earth and stone works. At no period has this process of demolition gone on in Scotland more rapidly and ruthlessly than during the last fifty or a hundred years. That tide of agricultural improvement which has passed over the country, has, in its utilitarian course, swept away—sometimes inevitably, often most needlessly—the aggers and ditches of ancient camps, sepulchral barrows and mounds, stone circles and cairns, earth-raths, and various other objects of deep antiquarian interest. Indeed, the chief antiquarian remains of this description which have been left on the surface of our soil are to be found on our mountain-tops, on our moors, or in our woods, where the very sterility or inaccessibility of the spot, or the kind protection and sympathy of the old forest-trees, have saved them, for a time at least, from reckless ruin and annihilation. Some of the antiquarian memorials that I allude to would have endured for centuries to come, had it not been for human interference and devastation.

For, in the demolition of these works of archaic man, the hand of man has too generally proved both a busier and a less scrupulous agent than the hand of time.

Railways have proved among the greatest, as well as the latest, of the agents of destruction. In our island various cherished antiquities have been often most unnecessarily swept away in constructing these race-courses for the daily rush and career of the iron horse. His rough and ponderous hoof, for example, has kicked down, at one extremity of a railway connected with Edinburgh (marvellously and righteously to the dispeace of the whole city), that fine old specimen of Scottish Second-Pointed architecture, the Trinity College Church; while, at the other extremity of the same line, it battered into fragments the old Castle of Berwick, a fort rich in martial and Border memories, and a building rendered interesting by the fact, that in connection with one of its turrets there was—at the command of Edward I. "the greatest of the Plantagenets," (as his latest biographer boastfully terms him)—constructed, some six centuries ago, a cage of iron and wood, in which he immured, with Bomba-like ferocity, for four weary years, a poor prisoner, and that prisoner a woman—the Countess of Buchan—whose frightful crime consisted in having assisted at the coronation of her liege sovereign, Robert the Bruce. In the construction of the Edinburgh and Glasgow Railway the line was driven, with annihilating effect, through the centre of the old and rich Roman Station on the Wall of Antoninus at Castlecary. Some years ago, as I passed along the line, I saw the farmer in the immediate neighbourhood of this station busily removing a harmless wall,—among the last, if not the very last remnants of Roman masonry in Scotland. The largest stone circle near the English Border—the Stonehenge or Avebury of the north of England—formerly stood near Shap. The stone avenues leading to it are said to have been nearly two miles in length. The engineer of the Carlisle and Lancaster railway carried his line right through the very centre of the ancient stone circle forming the head of the chief avenue, leaving a few of its huge stones standing out on the western side, where they may be still seen by the passing traveller about half a mile south of the Shap station. If the line had been laid only a few feet on either side, the wanton desecration and destruction of this fine archaic monument might have been readily saved. Railway engineers, however, and railway directors, care far more for mammon and money than for mounds and monoliths.

But other and older agents have overturned and uprooted the memorials transmitted down from ancient times, with as much wantonness as the railways. Towards the middle of the last century the Government of the day ordered many miles of the gigantic old Roman wall, which stretches across Northumberland and Cumberland, to be tossed over and pounded into road metal. About the same time a Scottish proprietor—with a Vandalism which cast a stigma on his order—pulled down that antique enigmatical building, "Arthur's Oven," in order to build, with its ashlar walls, a mill-dam across the Carron. At its next flood the indignant Carron carried away the mill-dam, and buried for ever in the depths of its own water-course those venerable stones which were begrudged any longer by the proprietor of the soil the few feet of ground which they had occupied for centuries on its banks.

In many parts of our country our old sepulchral cairns, hill-forts, castles, churches, and abbeys, have been most thoughtlessly and reprehensibly allowed, by those that chanced to be their proprietors for the time, to be used as mere quarries of ready stones for the building of villages and houses, and for the construction of field-dikes and drains. In the perpetration of this class of sad and discreditable desecrations, many parties are to blame. Such outrages have been practised by both landlord and tenant, by both State and Church; and I fear that the Presbyterian Church of Scotland is by no means free from much culpability in the matter. But let us, at the same time, rejoice that a better spirit is awakened on the whole question; and let us hope that our Scottish landlords will all speedily come to imitate, when required, the excellent example of Mr. Baillie, who, when some years ago he found that one of his tenants had pulled down and carried off, for building purposes, some portions of the walls of the four grand old burgs standing in Glenelg, in Inverness-shire, prosecuted the delinquent farmer before the sheriff-court of the county, and forced him to restore and replace *in situ*, as far as possible, and at his own expense, all the stones which he had removed.

Almost all the primæval stone circles and cromlechs which existed in the middle and southern districts of Scotland have been cast down and removed. The only two cromlechs in the Lothians, the stones of which have not been removed, are at Ratho and Kipps; and though the stones have been wantonly pulled down, they could readily be restored, and certainly deserve to be so. In 1813 the cromlech at Kipps was seen by Sir John Dalzell still standing upright. In describing it, in the beginning of the last century, Sir Robert Sibbald states that near this Kipps cromlech was a circle of stones, with a large stone or two in the middle; and he adds, "many such may be seen all over the country." They have all disappeared; and latterly the stones of the Kipps circle have been themselves removed and broken up, to build, apparently, some neighbouring field-walls, though there was abundance of stones in the vicinity equally well suited for the purpose.

Among the most valuable of our ancient Scottish monuments are certainly our Sculptured Stones. Most of them, however, and some even in late times, have been sadly mutilated and destroyed, to a greater or less degree, by human hands, and converted to the most base uses. The stone at Hilton of Cadboll, remarkable for its elaborate sculpture and ornamental tracery, has had one of its sides smoothed and obliterated in order that a modern inscription might be cut upon it to commemorate "Alexander Duff and His Thrie Wives." The beautiful sculptured stone of Golspie has been desecrated in the same way. Only two of these ancient sculptured stones are known south of the Forth. One of them has been preserved by having been used as a window-lintel in the church of Abercorn—the venerable episcopal see, in the seventh century, of Trumwine, the Bishop of the Picts. The other serves the purpose of a foot-bridge within a hundred yards of the spot where we are met; and it is to be hoped that its proprietors will allow this ancient stone to be soon removed from its present ignominious situation to an honoured place in our Museum. I saw, during last autumn, in Anglesey, a stone bearing a very ancient Romano-British legend, officiating as one of the posts of a park gate—a situation in which several such inscribed stones have been found. Still more lately, I was informed of the large central monolith in a stone circle, not far from the Scottish border, having been thrown down and split up into seven pairs of field gate-posts.

"Standing-stones"—the old names of which gave their appellations to the very manors on which they stood—have been repeatedly demolished in Scotland. An obelisk of thirteen feet in height, and imparting its name to a landed estate in Kincardineshire, was recently thrown down; and a large monolith, which lent its old, venerable name to a property and mansion within three or four miles of Edinburgh, was, within the memory of some living witnesses, uprooted and totally demolished when the direction of the turnpike road in its neighbourhood happened to be altered.

A healthier and finer feeling in regard to the propriety of preserving such national antiquities as I have referred to, subsists, I believe, in the heart of the general public of Scotland, than perhaps those who are their superiors in riches and rank generally give them credit for. Within this century the standing-stones of Stennis in Orkney were attacked, and two or three of them overthrown by an iconoclast; but the people in the neighbourhood resented and arrested the attempt by threatening to set fire to the house and corn of the barbaric aggressor. After the passing of the Parliamentary Reform Bill, during a keen contest for the representation of a large Scottish county, there was successfully urged in the public journals against one of the candidates, the damaging fact that one of his forefathers had deliberately committed one of the gross acts of barbarism which I have already specified, in the needless destruction, in a distant part of Scotland, of one of the smallest but most interesting of Scottish antiquarian relics; and the voters at the polling-booths showed that they deemed a family, however rich and estimable, unfit to be intrusted with the parliamentary guardianship of the county, which had outraged public feeling by wantonly pulling down one of the oldest stone memorials in the kingdom.

In the name of this Society, and in the name of my fellow-countrymen generally, I here solemnly protest against the perpetration of any more acts of useless and churlish Vandalism, in the needless destruction and removal of our Scottish antiquarian remains. The hearts of all leal Scotsmen, overflowing as they do with a love of their native land, must ever deplore the unnecessary demolition of all such early relics and monuments as can in any degree contribute to the recovery and restoration of the past history of our country and of our ancestors. These ancient relics and monuments are truly, in one strong sense, national property; for historically they belong to Scotland and to Scotsmen in general, more than they belong to the individual proprietors upon whose ground they accidentally happen to be placed. There is an Act of Parliament against the wilful defacing and demolition of public monuments; and, perhaps the Kilkenny Archæological Association were right when they threatened to indite with the penalties of "misdemeanour" under that statute, any person who should wantonly and needlessly destroy the old monumental and architectural relics of his country. Many of these relics might have brought only a small price indeed in the money-market, while yet they were of a national and historical value which it would be difficult to estimate. For, when once swept away, their full replacement is impossible. They cannot be purchased back with gold. Their deliberate and ruthless annihilation is, in truth, so far the annihilation of the ancient records of the kingdom. If any member of any ancient family among us needlessly destroyed some of the olden records of that one family, how bitterly, and how justly too, would he be denounced and despised by its members? But assuredly antiquarian monuments, as the olden records of a whole realm, are infinitely more valuable than the records of any individual family in that realm. Let us fondly hope and trust that a proper spirit of patriotism—that every feeling of good, generous, and gentlemanly taste—will insure and hallow the future consecration of all such Scottish antiquities as still remain—small fragments only though they may be of the antiquarian treasures that once existed in our land.

Time, like the Sibyl, who offered her nine books of destiny to the Roman king, has been destroying, century after century, one after another of the rich volumes of antiquities which she formerly tendered to the keeping of our Scottish fathers. But though, unhappily, our predecessors, like King Tarquin, rejected and scorned the rich antiquarian treasures which existed in their days, let us not now, on that account, despise or decline to secure the three books of them that still perchance remain. On the contrary,—like the priests appointed by the Roman authorities to preserve and study the Sibylline records which had escaped destruction,—let this Society carefully guard and cherish those antiquities of our country which yet exist, and let them strive to teach themselves and their successors to decipher and interpret aright the strange things and thoughts that are written on those Sibylline leaves of Scottish

Archæology which Fate has still spared for them. Working earnestly, faithfully, and lovingly in this spirit, let us not despair that, as the science of Archæology gradually grows and evolves, this Society may yet, in full truth, restore Scotland to antiquity, and antiquity to Scotland.

ON AN OLD STONE-ROOFED CELL OR ORATORY IN THE ISLAND OF INCHCOLM.[1]

AMONG the islands scattered along the Firth of Forth, one of the most interesting is the ancient Aemonia, Emona, St. Columba's Isle, or St. Colme's Inch—the modern Inchcolm. The island is not large, being little more than half-a-mile in length, and about a hundred and fifty yards across at its broadest part. At either extremity it is elevated and rocky; while in its intermediate portion it is more level, though still very rough and irregular, and at one point—a little to the east of the old monastic buildings—it becomes so flat and narrow that at high tides the waters of the Forth meet over it. Inchcolm lies nearly six miles north-west from the harbour of Granton, or is about eight or nine miles distant from Edinburgh; and of the many beautiful spots in the vicinity of the Scottish metropolis, there is perhaps none which surpasses this little island in the charming and picturesque character of the views that are obtained in various directions from it.

Though small in its geographical dimensions, Inchcolm is rich in historical and archæological associations. In proof of this remark, I might adduce various facts to show that it has been at one

[1] From the *Proceedings of the Society of Antiquaries of Scotland*, vol. ii. part iii.

time a favoured seat of learning, as when, upwards of four hundred years ago, the Scottish historian, Walter Bower, the Abbot of its Monastery, wrote there his contributions to the ancient history of Scotland;[1] and at other times the seat of war, as when it was pillaged at different periods by the English, during the course of the fourteenth, fifteenth, and sixteenth centuries.[2] For ages it was the

[1] These contributions by the "Abbas Aemoniæ Insulæ" are alluded to by Boece, who wrote nearly a century afterwards, as one of the works upon which he founded his own *Scotorum Historiæ*.—(See his *Praefatio*, p. 2; and Innes' *Critical Essay on the Ancient Inhabitants of Scotland*, vol. i., pp. 218 and 228.) Bower, in a versified colophon, claims the merit of having completed eleven out of the sixteen books composing the *Scotichronicon* lib. xvi. cap. 39):—

> "Quinque libros Fordun, undenos auctor arabat,
> Sic tibi clarescit sunt sedecim numero,
> Ergo pro precibus, petimus te, lector eorum," etc.

[2] See *Scotichronicon*, lib. xiii. cap. 34 and 37; lib. xiv. cap. 38, etc. In 1547 the Duke of Somerset, after the battle of Pinkie, seized upon Inchcolm as a post commanding "vtterly ye whole vse of the Fryth it self, with all the hauens uppon it," and sent as "elect Abbot, by God's sufferance, of the monastery of Sainct Coomes Ins, Sir Jhon Luttrell, knight, with C. hakbutters and l. pionerš, to kepe his house and land thear, and ii. rowe barkes, well furnished with municion, and lxx. mariners to kepe his waters, whereby (naively remarks Patten) it is thought he shall soon becum a prelate of great power. The perfytnes of his religion is not alwaies to tarry at home, but sumetime to rowe out abrode a visitacion; and when he goithe, I haue hard say he taketh alweyes his sumners in barke with hym, which ar very open mouthed, and neuer talk but they are harde a mile of, so that either for loove of his blessynges, or feare of his cursinges, he is lyke to be soouveraigne ouer most of his neighbours."—(See Patten's *Account of the late Expedicion in Scotlande*, dating "out of the parsonage of S. Mary Hill, London," in Sir John Dalyell's *Fragments of Scottish History*, pp. 79 and 81.) In Abbot Bower's time, the island seems to have been provided with some means of defence against these English attacks; for, in the *Scotichronicon*, in in-

site of a monastic institution and the habitation of numerous monks;[1] and at the beginning of the present century it was temporarily degraded to the site of a military fort, and the habitation of a corps of artillery.[2] During the plagues and epidemics of the sixteenth and seventeenth centuries, it formed sometimes a lazaretto for the suspected and diseased;[3] and during the reign of James I. it was used as a state-prison for the daughter of the Earl of Ross and the mother of the Lord of the Isles[4]—"a mannish, implacable woman," as Drummond of Hawthornden ungallantly terms her;[5] while fifty years later, when Patrick Graham, Archbishop of St. Andrews, was "decernit ane heretique, scismatike, symoniak, and declarit cursit, and condamnit to perpetuall presoun," he was,

cidentally speaking of the return of the Abbot and his canons in October 1421 from the mainland to the island, it is stated that they dared not, in the summer and autumn, live on the island for fear of the English, for, it is added, the monastery at that time was not fortified as it is now, "non enim erant tunc, quales ut nunc, in monasterio munitiones" (lib. xv. cap. 38).

[1] Iona itself has not an air of stiller solitude. Here, within view of the gay capital, and with half the riches of the Scotland of earlier days spread around them, the brethren might look forth from their secure retreat on that busy ambitious world, from which, though close at hand, they were effectually severed."—(Billings' *Baronial and Ecclesiastical Antiquities of Scotland*, vol. iii. Note on Inchcolm.)

[2] Alexander Campbell, in his *Journey through North Britain* (1802), after speaking of a fort in the east part of Inchcolm having a corps of artillery stationed on it, adds, "so that in lieu of the pious orisons of holy monks, the orgies of lesser deities are celebrated here by the sons of Mars," etc., vol. ii. p. 69.

[3] See MS. Records of the Privy Council of Scotland, 23d September 1564, etc.

[4] Bellenden's translation of Boece's *History of Scotland*, vol. ii. p. 500.

[5] *Works* of William Drummond, Edinburgh, 1711, p. 7.

for this last purpose, "first transportit to St. Colmes Insche."[1] Punishments more dark and dire than mere transportation to, and imprisonment upon Inchcolm, have perhaps taken place within the bounds of the island, if we do not altogether misinterpret the history of "a human skeleton standing upright," found several years ago immured and built up within the old ecclesiastic walls.[2] Nor is this eastern Iona, as patronised and protected by St. Columba,—and, at one period of his mission to the Picts and Scots, his own alleged dwelling-place,[3]—devoid in its history of the usual amount of old monkish miracles and legends. The Scotichronicon contains long and elaborate details of several of them. When, in 1412, the Earl of Douglas thrice essayed to sail out to sea, and was thrice driven back by adverse gales, he at last made a pilgrimage to the holy isle of Aemonia, presented an offering to Columba, and forthwith the Saint sped him with fair winds to Flanders and home again.[4] When, towards the winter of 1421, a boat was sent on a Sunday (die Dominica) to bring off to the monastery from the mainland some house provisions and barrels of beer brewed at Bernhill (in barellis cerevisiam apud Bernhill brasiatam), and the crew, exhilarated with liquor (alacres et potosi), hoisted, on their return, a sail, and upset the barge, Sir Peter the Canon,—who, with five others, was thrown into the water,—fervently and unceasingly invoked the aid of Columba, and the Saint appeared in person to him, and kept Sir Peter afloat for an hour and a half by the help of a

[1] Bishop Lesley's *History of Scotland*, p. 42.

[2] See General Hutton's MSS. in the Advocates' Library, as quoted in Billings' *Ecclesiastical Antiquities, loc. cit.*

[3] See his Life in Colgan's *Trias Thaumaturga*, vol. ii. p. 466.

[4] *Scotichronicon*, lib. xv. cap. 23.

truss of tow (adminiculo cujusdam stupæ), till the boat of Portevin picked up him and two others.[1] When, in 1385, the crew of an English vessel (quidam filii Belial) sacrilegiously robbed the island, and tried to burn the church, St. Columba, in answer to the earnest prayers of those who, on the neighbouring shore, saw the danger of the sacred edifice, suddenly shifted round the wind and quenched the flames, while the chief of the incendiaries was, within a few hours afterwards, struck with madness, and forty of his comrades drowned.[2] When, in 1335, an English fleet ravaged the shores of the Forth, and one of their largest ships was carrying off from Inchcolm an image of Columba* and a store of ecclesiastical plunder, there sprung up such a furious tempest around the vessel immediately after she set sail, that she drifted helplessly and hopelessly towards the neighbouring island of Inchkeith, and was threatened with destruction on the rocks there till the crew implored pardon of Columba, vowed to him restitution of their spoils, and a suitable offering of gold and silver, and then they instantly and unexpectedly were lodged safe in port (et statim in tranquillo portu insperate ducebantur).[3] When, in 1336, some English pirates

[1] *Scotichronicon*, lib. xv. cap. 38. [2] *Ibid.* lib. xv. cap. 48.

[3] *Ibid.* lib. xiii. cap. 34. When, in 1355, the navy of King Edward came up the Forth, and "spulyeit" Whitekirk, in East Lothian, still more

* [Images, or statues in wood, of the founders or patrons of churches of the sixth and seventh centuries, were common in Ireland, and no doubt in the Gaelic portion of Scotland. Some of these "images" are still preserved in islands on the west coast of Ireland. "St. Barr's wooden image" was preserved in his church in the island of Barray.—See Martin's *Western Isles of Scotland*, pp. 92, 93. But Macaulay, in his *History of St. Kilda*, p. 75, says, that this was an image of St. Brandan, to whom the church was consecrated. —P.]

robbed the church at Dollar—which had been some time previously repaired and richly decorated by an Abbot of Acmonia—and while they were, with their sacrilegious booty, sailing triumphantly, and with music on board, down the Forth, under a favouring and gentle west wind, in the twinkling of an eye (non solum subito sed in ictu oculi), and exactly opposite the abbey of Inchcolm, the ship sank to the bottom like a stone. Hence, adds the writer of this miracle in the *Scotichronicon,*—and no doubt that writer was the Abbot Walter Bower,—in consequence of these marked retaliating propensities of St. Columba, his vengeance against all who trespassed against him became proverbial in England; and instead of calling him, as his name seems to have been usually pronounced at the time, St. Căllum or St. Colām, he was commonly known among them as *St. Quhalme* ("et ideo, ut non reticeam quid de eo dicatur, apud eos vulgariter *Sanct Quhalme* nuncupatur" [1]).

But without dwelling on these and other well-known facts and fictions in the history of Inchcolm, let me state,—for the statement has, as we shall afterwards see, some bearing upon the more immediate object of this notice,—that this island is one of the few spots in the vicinity of Edinburgh that has been rendered classical by the pen of Shakspeare. In the second scene of the opening act of the tragedy of Macbeth, the Thane of Ross comes as a hurried messenger from the field of battle to King Duncan, and reports

summary vengeance was taken upon such sacrilege. For "trueth is (says Bellenden) ane Inglisman spulyeit all the ornamentis that was on the image of our Lady in the Quhite Kirk; and incontinent the crucifix fel doun on his head, and dang out his harnis."—(Bellenden's *Translation of Hector Boece's Croniklis*, lib. xv. c. 14; vol. ii. p. 446.)

[1] *Scotichronicon*, lib. xiii. cap. 37.

that Duncan's own rebellious subjects and the invading Scandinavians had both been so completely defeated by his generals, Macbeth and Banquo, that the Norwegians craved for peace :—

> "Sueno, the Norways' King, craves composition ;
> Nor would we deign him burial of his men
> Till he disbursed, at Saint Colmes Inch,
> Ten thousand dollars to our general use."

Inchcolm is the only island of the east coast of Scotland which derives its distinctive designation from the great Scottish saint. But more than one island on our western shores bears the name of St. Columba ; as, for example, St. Colme's Isle, in Loch Erisort, and St. Colm's Isle in the Minch, in the Lewis ; the island of Kolmbkill, at the head of Loch Arkeg, in Inverness-shire ; Eilean Colm, in the parish of Tongue ;[1] and, above all, Icolmkill, or Iona itself, the original seat and subsequent great centre of the ecclesiastic power of St. Columba and his successors.* An esteemed antiquarian friend, to whom I lately mentioned the preceding refer-

[1] See George Chalmers' *Caledonia*, vol. i. p. 320.

* ["Within the bay call'd *Loch-Colmkill*, three miles further south, lies *Lough Erisort*, which hath an anchoring-place on the south and north."—Martin, p. 4. "The names of the churches in Lewis Isles, and the saints to whom they were dedicated, are St. *Columbkil's*, in the island of that name," etc.—*Ibid.* p. 27. I suspect that all the churches founded by Columba bore *anciently* the name of Columbkill. Bede tells that the saint bore the united name of Columbkill.—*Hist. Ec.* v. 9 ; and all the churches founded by him in Ireland, or places called after him, are, I think, invariably so designated. Thus also the lake near Mugstot, in Skye, now drained, and on the island of which the most undoubted remains of a monastic establishment of Columb's time still exist, was called Lough Columbkill, and the island Inch Columbkill.—P.]

ence to Inchcolm by Shakspeare, at once maintained that the St. Colme's Isle in Macbeth was Iona. Indeed, some of the modern editors[1] of Shakspeare, carried away by the same view, have printed the line which I have quoted thus :—

> "Till he disbursed, at Saint Colme's-kill Isle,"

instead of "Saint Colmes ynch," as the old folio edition prints it. But there is no doubt whatever about the reading, nor that the island mentioned in Macbeth is Inchcolm in the Firth of Forth. For the site of the defeat of the Norwegian host was in the adjoining mainland of Fife, as the Thane of Ross tells the Scotch king that, to report his victory, he had come from the seat of war—

> "from Fife,
> Where the Norwegian banners flout the sky."

The reference to Inchcolm by Shakspeare becomes more interesting when we follow the poet to the original historical foundations upon which he built his wondrous tragedy. It is well known that Shakspeare derived the incidents for his story of Macbeth from that translation of Hector Boece's *Chronicles of Scotland*, which was published in England by Raphael Holinshed in 1577. In these Chronicles, Holinshed, or rather Hector Boece, after describing the reputed poisoning, with the juice of belladonna, of Sueno and his army, and their subsequent almost complete destruction, adds, that shortly afterwards, and indeed while the Scots were still celebrating this equivocal conquest, another Danish host landed at Kinghorn. The fate of this second army is described by Holinshed in the following words :—

[1] See, for example, the notes on this passage in the editions of Steevens and Malone.

"The Scots hauing woone so notable a victorie, after they had gathered and diuided the spoile of the field, caused solemne processions to be made in all places of the realme, and thanks to be giuen to almightie God, that had sent them so faire a day ouer their enimies. But whilest the people were thus at their processions, woord was brought that a new fleet of Danes was arriued at Kingcorne, sent thither by Canute, King of England, in reuenge of his brother Suenos ouerthrow. To resist these enimies, which were alreadie landed, and busie in spoiling the countrie, Makbeth and Banquho were sent with the Kings authoritie, who hauing with them a conuenient power, incountred the enimies, slue part of them, and chased the other to their ships. They that escaped and got once to their ships, obteined of Makbeth for a great summe of gold, that such of their friends as were slaine at this last bickering, might be buried in Saint Colmes Inch. In memorie whereof, manie old sepultures are yet in the said Inch, there to be seene grauen with the armes of the Danes, as the maner of burieng noble men still is, and hieretofore hath beene vsed. A peace was also concluded at the same time betwixt the Danes and Scotishmen, ratified (as some haue written) in this wise: that from thencefoorth the Danes should neuer come into Scotland to make anie warres against the Scots by anie maner of meanes. And these were the warres that Duncane had with forren enimies, in the seuenth yiere of his reigne."[1]

To this account of Holinshed, as bearing upon the question of the St. Colme's Isle alluded to by Shakspeare, it is only necessary to add one remark:—Certainly the western Iona, with its nine

[1] Holinshed's *Chronicles*, vol. v. p. 268.

separate cemeteries, could readily afford fit burial-space for the slain Danes ; but it is impossible to believe that the defeated and dejected Danish army would or could carry the dead and decomposing bodies of their chiefs to that remote place of sepulture. And, supposing that the dead bodies had been embalmed, then it would have been easier to carry them back to the Danish territories in England, or even across the German Ocean to Denmark itself, than round by the Pentland Firth to the distant western island of Icolmkill. On the other hand, that St. Colme's Inch, in the Firth of Forth, is the island alluded to, is, as I have already said, perfectly certain, from its propinquity to the seat of war, and the point of landing of the new Scandinavian host, namely, Kinghorn ; the old town of Wester Kinghorn lying only about three or four miles below Inchcolm, and the present town of the same name, or Eastern Kinghorn, being placed about a couple of miles further down the coast.

We might here have adduced another incontrovertible argument in favour of this view by appealing to the statement, given in the above quotation, of the existence on Inchcolm, in Boece's time, of Danish sepulchral monuments, provided we felt assured that this statement was in itself perfectly correct. But before adopting it as such, it is necessary to remember that Boece describes the sculptured crosses and stones at Camustane and Aberlemno,[1] in Forfarshire, as monuments of a Danish character also ; and whatever may have been the origin and objects of these mysteries in Scottish archæology,—our old and numerous Sculptured Stones, with their strange enigmatical symbols,—we are at least certain

[1] *Scotorum Historiæ*, lib. xi. f. 225, 251.

that they are not Danish either in their source or design, as no sculptured stones with these peculiar symbols exist in Denmark itself. That Inchcolm contained one or more of those sculptured stones, is proved by a small fragment that still remains, and which was detected a few years ago about the garden-wall. A drawing of it has been already published by Mr. Stuart.[1] (See woodcut, Fig. 1.) In the quotation which

Fig. 1. Sculptured Stone, Inchcolm.

I have given from Holinshed's Chronicles, the "old sepultures there (on Inchcolm) to be seene grauen with the armes of the Danes," are spoken of as "manie" in number.* Bellenden uses similar language: "Thir Danes" (he writes) "that fled to thair schippis, gaif gret sowmes of gold to Makbeth to suffer thair freindis that war slane at his jeoperd to be buryit in Sanct Colmes Inche. In memory heirof, *mony* auld sepulturis ar yit in the said Inche, gravin with armis of Danis."[2] In translating this passage from Boece, both Holinshed and Bellenden overstate, in some degree, the words of their original author. Boece speaks of the Danish monuments still existing on Inchcolm in his day, or about the year

[1] See his great work on the *Sculptured Stones of Scotland*, plate cxxv. p. 39.

[2] Bellenden's *Translation of Boece's Croniklis of Scotland*, lib. xii. 2, vol. ii. p. 258.

---

* [I do not believe that there is a single example of armorial bearings to be found either in Scotland or Ireland of an earlier date than the close of the twelfth century.—P.]

1525, as plural in number, but without speaking of them as many. After stating that the Danes purchased the right of sepulture for their slain chiefs (nobiles) "in Emonia insula, loco sacro," he adds, "extant et hac ætate notissima Danorum monumenta, lapidibusque insculpta eorum insignia."[1] For a long period past only one so-called Danish monument has existed on Inchcolm, and is still to be seen there. It is a single recumbent block of stone above five feet long, about a foot broad, and one foot nine inches in depth, having a rude sculptured figure on its upper surface. In his *History of Fife*, published in 1710, Sir Robert Sibbald has both drawn and described it. "It is (says he) made like a coffin, and very fierce and grim faces are done on both the ends of it. Upon the middle stone which supports it, there is the figure of a man holding a spear in his hand."[2] He might have added that on the corresponding middle part of the opposite side there is sculptured a rude cross; but both the cross and "man holding a spear" are cut on

Fig. 2. Danish Monument.

the single block of stone forming the monument, and not, as he represents, on a separate supporting stone. Pennant, in his *Tour through Scotland in* 1772, tells us that this "Danish monument" "lies in the south-east [south-west] side of the building (or monas-

[1] *Scotorum Historiæ* (1526), lib. xii. p. 257.

[2] *History of Fife and Kinross*, p. 35.

tery), on a rising ground. It is (he adds) of a rigid form, and the surface ornamented with scale-like figures. At each end is the representation of a human head."[1]* In its existing defaced form,[2]

[1] *A Tour in Scotland*, part ii. p. 210. See also Grose's *Antiquities of Scotland* (1797), vol. ii. p. 135.

[2] In the *Buik of the Croniclis of Scotland*, or metrical version of the History of Hector Boece, by William Stewart, lately published under the authority of the Master of the Rolls, and edited by Mr. Turnbull, there is a description of the Danish monument on Inchcolm from the personal observation of the translator; and we know that this metrical translation was finished by the year 1535. The description is interesting, not only from being in this way a personal observation, but also as showing that, at the above date, the recumbent sculptured "greit stane," mentioned in the text, was regarded as a monument of the Danish leader, and that there stood beside it a Stone Cross, which has since unfortunately disappeared. After speaking of the burial of the Danes—

> Into an yle callit Emonia,
> Sanct Colmis hecht now callit is this da,

and the great quantity of human bones still existing there, he adds in proof—

> As *I myself quhilk has bene thair and sene.*
> Ane croce of stane thair standis on ane grene,
> Middis the feild quhair that they la ilk one,
> Besyde the croce thair lyis ane greit stane;
> Under the stane, in middis of the plane,
> Their chiftane lyis quhilk in the feild was slane.

(See vol. ii. p. 635). Within the last few months there has been discovered by Mr. Crichton another sculptured stone on Inchcolm. But the character of the sculptures on it is still uncertain, as the stone is in a dark corner, the exposed portion of it forming the ceiling of the staircase of the Tower, and the remainder of the stone being built into, and buried in the wall. The sculptures are greatly weather-worn, and the stone itself had been used in the original building of the Tower. The Tower of St. Mary's Church, or of the

* [I feel quite satisfied that this monumental stone is of a much earlier date than the thirteenth century, and that it is most probably a Danish or Dano-Scottish monument.—P.]

the sculpture has certainly much more the appearance of a recumbent human figure, with a head at one end and the feet at the other, than with a human head at either extremity. The present

so-called Cathedral at Iona, is known to have been erected early in the thirteenth century. Mr. Huband Smith, who believes the Tower of the Cathedral in Iona, and perhaps the larger portion of the nave and aisles, to be "probably the erection of the twelfth and next succeeding century," found, in 1844, on the abacus of one of the supporting columns, the inscription "DONALDUS OBROLCHAN FECIT HOC OPUS;" and already this inscription has been broken and mutilated.—(See Ulster *Journal of Archæology*, vol. i. p. 86.) The obit of a person of this name, and probably of this builder, occurs, as Dr. Reeves has shown, in the *Annals of Ulster* in 1203, and in the *Annals of the Four Masters* in 1202; and Dr. Reeves considers the Church or Cathedral at Iona as "an edifice of the early part of the thirteenth century."—(*Life of Columba*, pp. 411 and 416.) But the Tower of the Church of Inchcolm is so similar in its architectural forms and details to that of Icolmkill, that it is evidently a structure nearly, if not entirely, of the same age; and the new choir (novum chorum) built to the church in 1265 (see *Scotichronicon*, lib. x. c. 20) is apparently, as seen by its remaining masonic connections, posterior in age to the Tower upon which it abuts. Hence we are, perhaps, fairly entitled to infer that this sculptured stone thus incidentally used in the construction of the Tower on Inchcolm, existed on the island long, at least, before the thirteenth century, as by that time it was already very weather-worn, and consequently old.*

---

* [I, too, consider this church to be of the early part of the thirteenth century. Parts of it, however, I believe to be of the twelfth century. I allude particularly to that portion on one of the columns of which the name of the builder appears, and who, I have little doubt, was the eminent person whose death—1202—is recorded by the Annalists. Pinkerton, vol. ii. p. 258, is in error in supposing any portion of the church to be of the eleventh century. The family of the O'Brolchans were of distinguished rank in the county of Derry, and intimately connected with the churches there. See my notices of them in the *Ordnance Memoir of the Parish of Temple More*, pp. 21, 22, 29. It may be worthy of remark that this family of O'Brolchain, or a

condition of the monument is faithfully given in the accompanying woodcut, which, like most of the other woodcuts in this little essay, have been copied from sketches made by the masterly pencil of my esteemed friend, Mr. James Drummond, R.S.A.

It is well known that, about a century after the occurrence of these Danish wars, and of the alleged burial of the Danish chiefs on Inchcolm,—or in the first half of the thirteenth* century,—there was founded on this island, by Alexander I., a monastery, which from time to time was greatly enlarged, and well endowed. The monastic buildings remaining on Inchcolm at the present day are of very various dates, and still so extensive that their oblong light-grey mass, surmounted by a tall square central tower, forms a striking object in the distance, as seen in the summer morning light from the higher streets and houses of Edinburgh, and from the neighbouring shores of the Firth of Forth. These monastic buildings have been fortunately protected and preserved by their insular situation,—not from the silent and wasting touch of time, but from the more ruthless and destructive hand of man. The stone-roofed octagonal chapter-house is one of the most beautiful and perfect in Scotland; and the abbot's house, the cloisters, re-

---

branch of it, appear to have been eminent, hereditarily, after the Irish usage, as architects or builders. At the year 1029 the *Annals of Ulster* record the death of Maolbride O'Brolchan, "*chief mason* of Ireland." And at the year 1097, the death of Maelbrighde *Mac-an-tsaeir* (son of the mason) O'Brolchan. And, lastly, we have the name of Donald O'Brolchan as the architect of the great church at Iona. But if this Donald be the person whose death is recorded in the *Annals* as "a noble senior" in 1202, that part of the building in which the inscription is found must be surely of the twelfth century; and the style of its architecture supports that conclusion.—P.]

* [Twelfth.—P.]

fectory, etc., are still comparatively entire. But the object of the present communication is not to describe the well-known conventual ruins on the island, but to direct the attention of the Society

Fig 3. Inchcolm.

to a small building, isolated, and standing at a little distance from the remains of the monastery, and which, I am inclined to believe, is of an older date, and of an earlier age, than any part of the monastery itself.

The small building, cell, oratory, or chapel, to which I allude, forms now, with its south side, a portion of the line of the north wall of the present garden, and is in a very ruinous state ; but its more characteristic and original features can still be accurately made out.

The building is of the quadrangular figure of the oldest and smallest Irish churches and oratories. But its form is very irregular, partly in consequence of the extremely sloping nature of the ground on which it is built, and partly perhaps to accommodate it in position to three large and immovable masses of trap that lie on either side of it, and one of which masses is incorporated into its south-west angle. It is thus deeper on its north than on its south side; and much deeper at its eastern than at its western end. Further, its remaining eastern gable is set at an oblique angle to the side walls, while both the side walls themselves seem slightly curved or bent. Hence it happens, that whilst externally the total length of the north side of the building is 19 feet and a half, the total length of its south side is 21 feet and a half, or 2 feet more.

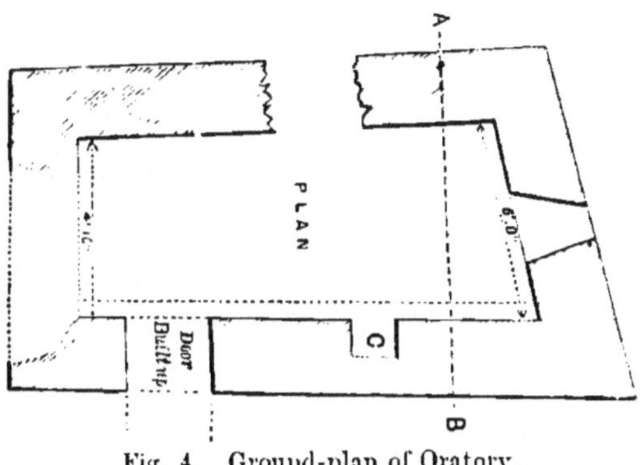

Fig. 4. Ground-plan of Oratory.

Internally, also, it gradually becomes narrower towards its western extremity; so that, whilst the breadth of the interior of the building is about 6 feet 3 inches at its eastern end, it is only 4 feet and

9 inches at its western end. Some of these peculiarities are shown in the accompanying ground-plan drawn by Mr. Brash (see woodcut, Fig. 4), in which the line A B represents the whole breadth of the building; A the north, and B the south wall of it. Unfortunately, as far as can be gathered amid the accumulated debris at the western part of the building, the gable at that end is almost destroyed, with the exception of the stones at its base; but, judging from the height of the vaulted roof, this gable probably did not measure externally above 8 feet, while the depth of the eastern gable, which is comparatively entire, is between 14 and 15 feet. The interior of the building has been originally, along its central line, about 16 feet in length; it is nearly 8 feet in height from the middle of the vaulted roof to the present floor; and the interior has an average breadth of about 5 feet. Internally the side walls are 5 feet in height from the ground to the spring of the arch or vault.

Three feet from the ground there is interiorly, in the south wall, a small four-sided recess,* 1 foot in breadth, and 15 inches in height and depth. (See C in ground plan, Fig. 4; and also Fig. 8.) In the same south-side wall, near the western gable, is an opening extending from the floor to the spring of the roof. It has apparently been the original door of the building; but as it is now built up by a layer of thin stone externally, and the soil of the garden has been heaped up against it and the whole south wall to the depth of several feet, it is difficult to make out its full relations and character. There is a peculiarity, however, about the head of

* [Square recesses or ambries of this kind are common in the most ancient Irish oratories.—P.]

this entrance which deserves special notice. The top of the doorway, as seen both from within and from without the building, is arched, but in two very different ways. When examined from within, the head of the doorway is found to be composed of stones laid in the form of a horizontal arch, the superincumbent stones on each side projecting more and more over each other to constitute its sides, and

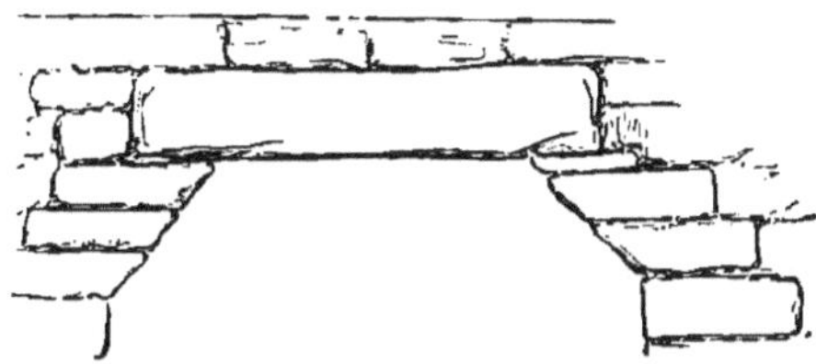

Fig. 5. Horizontal arch of the door, as seen from within the cell.

then a large, flat, horizontal stone closing the apex. (See woodcut, Fig. 5.) On the contrary, when examined from without, the top of the doorway is formed by stones laid in the usual form of the radiating arch, and roughly broken off, as if that arch at a former

Fig. 6. Semi-circular arch of the door as seen from without, the garden earth filling the doorway.

period had extended beyond the line of the wall. (See woodcut, Fig. 6.) This doorway, let me add, is 5 feet high, and on an average about 4 feet wide,* but it is 2 or 3 inches narrower at the top, or at the spring of the arch, than it is at the bottom.[1] The north side wall of the building is less perfect; as, in modern times, a large rude opening has been broken through as an entrance or door (see woodcut, Fig. 7, and ground-plan, Fig. 4), after the original door on the other side had become blocked up.

[1] When I first visited Inchcolm the ancient cell described in the present paper was the abode of one or two pigs; and on another occasion I found it inhabited by a cow. In consequence of the attention of the Earl of Moray (the proprietor of the island), and his active factor, Mr. Philipps, having been directed to the subject, all such desecration has been put an end to, and the whole building has been repaired in such a way as to retard its dilapidation. The plans required for its proper repair were kindly drawn out by my friend Mr. Brash of Cork, a most able architect and archæologist, who had performed on various occasions previously a similar duty in reference to the restoration of old ecclesiastical buildings in the south and west of Ireland. All these restorations preserve, as far as possible, in every respect the original characteristics of the building. In making these restorations, several points mentioned in the text as visible in the former dilapidated state of the building, are now of course covered up, such as the section of the arch of the roof, represented in woodcut, Fig. 9, etc. Other new points, not alluded to in the text, were cleared up and brought to light as the necessary repairs were proceeded with. The opening in the western part of the south wall of the building was found to be the undoubted original door of the cell; and when the earth accumulated up against it externally was cleared away, there was discovered, leading from this door to the south, and in the direction of the

[* The unusual breadth, 4 feet, of this doorway, is perhaps the only feature in the structure likely to excite a doubt of its early antiquity. I cannot remember ever having seen in any very ancient church or oratory in Ireland a doorway so wide. The widest doorway that I have met with is, I think, that of the great church at Glandelough, which is 3 feet 10 inches wide at its base. The usual width in doorways of small churches and oratories is from 2 feet to 2 feet 10 inches.—P.]

The eastern gable is still very entire, and contains a small window,* which, as measured outside, is 1 foot 11 inches in height,

well of the island, a built way or passage,† gently sloping upwards out of the cell, 4 feet in width, like the door itself, but becoming slightly wider when it reached the limit to which it has been as yet traced—viz., about 13 or 14 feet from the building. The built sides of this passage still stand about 3 or 4 feet in height; the lime used as cement in constructing these sides is apparently the same as that used in the construction of the walls of the cell itself; and, further, the passage has been coated over with the same dense plaster as that still seen adhering at different points to the interior of the oratory. It is impossible to fix the original height of the walls of this passage, but probably these walls were so high at one time, near the entrance at least into the oratory, as to be there arched over; for, as stated in the text, the stones composing the outer or external arch of the doorway offer that appearance of irregular fracturing which they would necessarily show if the archway had been originally continued forward, and subsequently broken across parallel with the line or face of the south side wall. It is perhaps not uninteresting here to add, that in Icolmkill a similar walled walk or entrance led into the small house or building of unknown antiquity, named the "Culdee's Cell." In the old *Statistical Account* (1795), this cell is described as "the foundation of a small circular house, upon a reclining plain. From

* [This window seems very ancient, and no mistake! Compare it with the window of the oratory near Kilmalkedar, in my *Towers*, p. 184. First edition.—P.]

† [This fact is, I think, very interesting and important as an evidence of the great antiquity of the building. Such built-passages are often found in Ireland connected with small churches and oratories of the sixth and seventh centuries, but never, to my knowledge, with any of a later age. They may, in fact, be considered as characteristic appendages, or accompanying features, to the ecclesiastical structures of those times. There is one at Rathmichael, near Dublin, where there is the butt of a round tower. I have seen many of them in various states of preservation, and I think all were about 4 feet both in breadth and height. They were, however, never arched, but roofed with large flags, laid horizontally, and their upper surface level with the surrounding ground.—P.]

and 10 inches in breadth. But the jambs of this window incline or splay internally, so as to form on the internal plane of the gable an opening 2 feet 3 inches in breadth.

Fig. 7. Eastern gable and north side of the building.

The squared sill stone of the window is one of the largest in the eastern gable. Its flat lintel stone projects externally in an angled or sharpened form beyond the plane of the gable, like a rude attempt at a moulding or architrave, but probably with the more

the door of the house a walk ascends to a small hillock, with the remains of a wall upon each side of the walk, which grows wider to the hillock."—(*Statistical Account of Scotland*, vol. xiv. p. 200.) At the old heremetical establishment of St. Fechin, on High Island, Connemara, there is "a covered passage, about 15 feet long and 3 wide," leading from the oratory to the supposed nearly circular, dome-roofed cell of the Abbot.—(Dr. Petrie's *Ecclesiastical Architecture*, p. 425.)

utilitarian object of preventing entrance of the common eastern showers into the interior of the cell. The thin single flat sandstones composing the jambs are each large enough to extend backwards the whole length of the interior splay of the window, and, from the marks upon them, have evidently been hammer-dressed.* Internally, in this eastern gable, there is placed below the window, and in continuation of its interior splay, a recess about 18 inches in depth, and of nearly the same breadth as the divergence of the jambs of the window. The broken base or floor of this recess is in the position of the altar-stone in some small early Irish chapels.

The accompanying sketch (see woodcut, Fig. 7) of the exterior of the eastern gable shows that the stones of which it is built have been prepared and dressed with sufficient care—especially those forming the angles—to entitle us to speak of it as presenting the type of rude ashlar-work. The stones composing it, particularly above the line of the window, are laid in pretty regular horizontal courses; lower down they are not by any means so equable in size. The masonry of the side walls is much less regular, and more of a ruble character. The walls are on an average about 3 feet in thickness.† The stones of which the building is composed are, with a few exceptions, almost all squared sandstone. The exceptions consist of some larger stones of trap or basalt, placed principally along the base of the walls. Both secondary trap and sandstone are found *in situ* among the rocks of the island. A roundish basalt stone, 2 feet long, forms a portion of the floor of the building at its

* After this sentence Dr. Petrie adds, "Good—very good."

† [This is a strong evidence in favour of the antiquity of the structure. —P.]

southern corner. At other points there is evidence of a well-laid earth floor. The whole interior of the building has been carefully plastered at one time. The surface of this plaster-covering of the walls, wherever it is left, is so dense and hard as to be scratched with difficulty. The lime used for building and cementing the walls, as shown in a part at the west end which has been lately exposed, contains oyster and other smaller sea-shells, and is as firm and hard as some forms of concrete.

I have reserved till the last a notice of one of the most remarkable architectural features in this little building, namely, its arched or vaulted stone roof,—the circumstance, no doubt, to which the whole structure owes its past durability and present existence.

Stone roofs are found in some old Irish buildings, formed on the principle of the horizontal arch, or by each layer of stone overlapping and projecting within the layer placed below it till a single stone closes the top. A remarkable example of this type of stone roof is presented by the ancient oratory of Gallerus in the county of Kerry; and stone roofs of the same construction covered most of the old beehive houses and variously shaped cloghans that formerly existed in considerable numbers in the western and southern districts of Ireland, and more sparsely on the western shores of Scotland. In the Inchcolm oratory the stone roof is constructed on another principle—on that, namely, of the radiating arch—a form of roof still seen in some early Irish oratories and churches, whose reputed date of building ranges from the sixth or seventh onward to the tenth or eleventh centuries.

The mode of construction of the stone roof of the Inchcolm cell is well displayed in the accidental section of it that has been made by the falling in of the western gable. One of Mr. Drummond's sketches (see woodcut, Fig. 9) represents the section as seen across the collection of flower-tipped rubbish and stones made by the

Fig. 8. Interior of the building, showing splayed window in eastern gable, recess in interior of south wall, vaulted roof, etc.

debris of the gable and some accumulated earth. The roof is constructed, first, of stones placed in the shape of a radiating arch; secondly, of a thin layer of lime and small stones placed over the outer surface of this arch; and, thirdly, the roof is finished by being covered externally with a layer of oblong, rhomboid stones, laid in regular courses from the top of the side walls onwards and upwards to the ridge of the building. This outer coating of squared stones is seen in the external surface of the roof to the left in one sketch (see woodcut, Fig. 9); but a more perfect and better pre-

Fig. 9. Exposed section of the arch of the vault.

served specimen of it exists immediately above the entrance-door, as shown in another of Mr. Drummond's drawings (see woodcut, Fig. 6).

The arch or vault of the roof has one peculiarity, perhaps worthy of notice (and seen in the preceding woodcut, Fig. 9). The central keystone of the arch has the form of a triangular wedge, or of the letter V, a type seen in other rude and primitive arches. Interiorly, a similar keystone line appears to run along the length of the vault, but not always perfectly straight; and the whole figure of the arch distinctly affects the pointed form.

Several years ago I first saw the building which I have described when visiting Inchcolm with Captain Thomas, Dr. Daniel Wilson, and some other friends, and its peculiar antique character and strong rude masonry struck all of us, for it seemed different in type from any of the other buildings around it. Last year I had an opportunity of visiting several of the oldest remaining Irish churches and oratories at Glendalough, Killaloe, Clanmacnoise, and elsewhere, and the features of some of them strongly recalled to my recollection the peculiarities of the old building in Inchcolm, and left on my mind a strong desire to re-inspect it. Later in the year Mr. Fraser and I visited Inchcolm in company with our greatest Scottish authority on such an ecclesiological question—Mr. Joseph Robertson. That visit confirmed us in the idea, first, that the small building in question was of a much more ancient type than any portion of the neighbouring monastery; and, secondly, that in form and construction it presented the principal architectural characters of the earliest and oldest Irish churches and oratories. More lately I had an opportunity of showing the various original sketches which Mr. Drummond had made for me of the building to the highest living authority on every question connected with early Irish and Scoto-Irish ecclesiastical architecture—namely, Dr. Petrie of Dublin; and before asking anything as to its site, etc., he at once pronounced the building to be "a Columbian cell."

The tradition, as told to our party by the cicerone on the island on my first visit, was, that this neglected outbuilding was the place in which "King Alexander lived for three days with the hermit of Inchcolm." There was nothing in the rude architecture and general character of the building to gainsay such a tradition, but the reverse; and, on the contrary, when we turn to the notice of a visit of Alexander I. to the island in 1123, as given by our earliest Scotch historians, their account of the little chapel or oratory which he found there perfectly applies to the building which I have been describing. In order to prove this, let me quote the history of Alexander's visit from the *Scotichronicon* of Fordun and Bower, the *Extracta e Cronicis Scocie*, and the *Scotorum Historia* of Hector Boece.[1]

The *Scotichronicon* contains the following account of King Alexander's adventure and temporary sojourn in Inchcolm :—

"About the year of our Lord 1123, under circumstances not less wonderful than miraculous, a monastery was founded on the island Aemonia, near Inverkeithing. For when the noble and most Christian Sovereign Alexander, first of this name, was, in pursuit of some state business, making a passage across the Queensferry, suddenly a tremendous storm arose, and the fierce south-west wind forced the vessel and sailors to make, for safety's sake, for the island of Aemonia, where at that time lived an island hermit (*eremita insulanus*), who, belonging to the service of St. Columba, devoted himself sedulously to his duties at a certain little chapel there (*ad quandam inibi capellulam*), content with such poor food as the milk of one cow and the shell and small sea fishes which he could collect. On the hermit's slender stores the king and his suite of companions, detained by the storm, gratefully lived for three consecutive days. But on the day before landing, when in very great danger from the sea, and tossed by the fury of the tempest, the king despaired of life, he vowed to the Saint, that if he should bring him and his companions safe to the island, he would leave

[1] See other similar notices of the visit of Alexander I. to Inchcolm in Buchanan's *Rerum Scoticarum Historia*, lib. vii. cap. 27 ; Leslæus *de Rebus Gestis Scotorum*, lib. vi. p. 219, etc.

on it such a memorial to his honour as would render it a future asylum and refuge to sailors and those that were shipwrecked. Therefore, it was decided on this occasion that he should found there a monastery of prebendaries, such as now exists; and this the more so, as he had always venerated St. Columba with special honour from his youth; and chiefly because his own parents were for several years childless and destitute of the solace of offspring, until, beseeching St. Columba with suppliant devotion, they gloriously obtained what they sought for so long a time with anxious desire. Hence the origin of the verse—

'M.C, ter, I. bis, et X literis à tempore Christi,
Aemon, tunc ab Alexandro fundata fuisti
Scotorum primo. Structorem Canonicorum
Transferat ex imo Deus hunc ad alta polorum.'"[1]

The preceding account of King Alexander's visit to Inchcolm, and his founding of the monastery there, occurs in the course of the fifth book (lib. v. cap. 37) of the *Scotichronicon*, without its being marked whether the passage itself exists in the original five books of Fordun, or in one of the additions made to them by the Abbot Walter Bower.[2] The first of these writers, John of Fordun, lived,

[1] Joannis de Fordun *Scotichronicon*, cum Supplementis et Continuatione Walteri Boweri Insulæ St. Columbæ Abbatis; cura Walteri Goodall (1759), vol. i. p. 286.

[2] My friend Mr. David Laing, with his usual kindness, has examined, with a view to this point, several manuscripts of the *Scotichronicon*, and has found that the account in that work of King Alexander's visit to Inchcolm is from the pen of Bower, and, as Mr. Laing adds in his note to me, "not the less curious and interesting on that account." In his original portion of the History, Fordun himself merely refers to the foundation of the Monastery of Inchcolm by Alexander.

it will be recollected, in the reigns of Robert II. and III., and wrote about 1380; while Walter Bower, the principal continuator of Fordun's history, was Abbot of Inchcolm from 1418 to the date of his death in 1449.

In the work known under the title of *Extracta e Variis Cronicis Scocie,*[1] there is an account of Alexander's fortuitous visit to Inchcolm, exactly similar to the above, but in an abridged form. Mr. Tytler, in his *History of Scotland,*[2] supposes the *Extracta* to have been written posterior to the time of Fordun, and prior to the date of Bower's *Continuation of the Scotichronicon,*—a conjecture which one or more passages in the work entirely disprove.[3] If the opinion of Mr. Tytler had been correct, it would have been important as a proof that the story of the royal adventure of Alexander upon Inchcolm was written by Fordun, and not by Bower, inasmuch as the two accounts in the *Scotichronicon* and in the *Extracta* are on this, as on most other points, very similar, the *Extracta* being merely somewhat curtailed. As evidence of this remark, let me here cite the original words of the *Extracta* :—

"Emonia insula seu monasterium, nunc Sancti Columbe de Emmonia, per dictum regem fundatur circa annum Domini millesimum vigesimum quartum miraculose. Nam cum idem nobilis rex transitum faciens per Passagium Regine, exorta tempestas valida, flante Africo, ratem cum naucleris, vix vita comite, compulit applicare ad insulam Emoniam, ubi tunc degebat quidam heremita insulanus, qui seruicio Sancti Columbe deditus, ad quam-

[1] *Extracta e Cronicis Scocie*, p. 66. [2] *History of Scotland*, vol. iii. p. 336.

[3] See Mr. Turnbull's Introductory Notice to the Abbotsford Club edition of the *Extracta*, p. xiv.

dam inibi capellulam tenui victu, utpote lacte vnius vacce et conchis ac pisiculis marinis contentatus, sedule se dedit, de quibus cibariis rex cum suis, tribus diebus, vento compellente, reficitur. Et quia Sanctum Columbam a juventute dilexit, in periculo maris, ut predicitur, positus, vouit se, si ad prefatam insulam veheretur incolumis, aliquid memoria dignum ibidem facere, et sic monasterium ibidem construxit canonicorum, et dotauit."[1]

I shall content myself with citing from our older Scottish historians one more account of Alexander's adventure upon Inchcolm—namely, that given by Hector Boece, Principal of King's College, Aberdeen, in his *Scotorum Historia*, a work written during the reign of James V., and first published in 1526. In this work, after alluding to the foundation of the Abbey of Scone, Boece proceeds to state that—(to quote the translation of the passage as given by Bellenden)—"Nocht long efter King Alexander come in Sanct Colmes Inche ; quhair he was constrainit, be violent tempest, to remane thre dayis, sustenand his life with skars fude, be ane heremit that dwelt in the said inche : in quhilk, he had ane litill chapell, dedicat in the honoure of Sanct Colme. Finaly, King Alexander, becaus his life was saiffit be this heremit, biggit ane Abbay of Chanonis regular, in the honour of Sanct Colme ; and dotat it with sindry landes and rentis, to sustene the abbot and convent thairof."[2]

As Bellénden's translation of Boece's work does not in this and other parts adhere by any means strictly to the author's original

[1] *Extracta e Cronicis Scocie*, p. 66.

[2] Boece's *History and Chronicles of Scotland*, translated by John Bellenden, book xii. chap. 15, vol. ii. p. 294.

context, I will add the account given by Boece in that historian's own words :[1]

"Nec ita multo post Fortheæ rex æstuarium trajiciens, coorta tempestate in Emoniam insulam appulsus descendit, repertoque Divi Columbæ *sacello,* viroque Eremita, triduo tempestatis vi permanere illic coactus est, exiguo sustentatus cibo, quem apud Eremitam quendam sacelli custodem reperiebat, nec tamen comitantium multitudini ulla ex parte sufficiente. Itaque eo periculo defunctus Divo Columbæ ædem vovit. Nec diu voto damnatus fuit, cœnobio paulo post Regularium, ordinis Divi Augustini extructo, agrisque atque redditibus ad sumptus eorum collatis."

That the very small and antique-looking edifice which I have described as still standing on Inchcolm is identically the little chapel or cell spoken of by Fordun and Boece as existing on the island at the time of Alexander's visit to it, upwards of seven centuries ago, is a matter admitting of great probability, but not of perfect legal proof. One or two irrecoverable links are wanting in the chain of evidence to make that proof complete ; and more particularly do we lack for this purpose any distinct allusions or notices among our mediæval annalists, of the existence or character of the building during these intervening seven centuries, except, indeed, we consider the notice of it which I have cited from the *Scotichronicon "ad quandam inibi capellulam,"* to be written by the hand of Walter Bower, and to have a reference to the little chapel as it existed and stood about the year 1430, when Bower wrote his additions to Fordun, while living and ruling on Inchcolm as Abbot of its Monastery.

[1] *Scotorum Historia*, lib. v. fol. cclxxii. First Paris Edition of 1526.

But various circumstances render it highly probable that the old stone-roofed cell still standing on the island is the ancient chapel or oratory in which the island hermit (*eremita insulanus*) lived and worshipped at the time of Alexander's royal but compulsory visit in 1123. I have already adduced in favour of this belief the very doubtful and imperfect evidence of tradition, and the fact that this little building itself is, in its whole architectural style and character, evidently far more rude, primitive, and ancient, than any of the extensive monastic structures existing on the island, and that have been erected from the time of Alexander downwards. In support of the same view there are other and still more valuable pieces of corroborative proof, which perhaps I may be here excused from now dwelling upon with a little more fullness and detail.

The existing half-ruinous cell answers, I would first venture to remark—and answers most fitly and perfectly—to the two characteristic appellations used respectively in the *Scotichronicon* and in the *Historiæ Scotorum*, to designate the cell or oratory of the Inchcolm anchorite at the time of King Alexander's three days' sojourn on the island. These two appellations we have already found in the preceding quotations to be *capellula* and *sacellum*. As applied to the small, rude, vaulted edifice to which I have endeavoured to draw the attention of the Society, both terms are strikingly significant. The word used by Fordun or Bower in the *Scotichronicon* to designate the oratory of the Inchcolm anchorite, namely "capellula," or little chapel, is very descriptive of a diminutive church or oratory, but at the same time very rare. Du Cange, in his learned glossary, only adduces one example of its employment. It occurs in the testament of Guido, Bishop of Auxerrè, in the thirteenth century (1270), who directs that "oratorium seu *capellulam* super sepulchrum dicti Robini construent." This passage further proves the similar signification of the two names of oratorium and capellula. The other appellation "sacellum," applied by Boece to the hermit's chapel, is a better known and more classical word than the capellula of the *Scotichronicon.* It is, as is well known, a diminutive from sacer, as tenellus is from tener, macellus from macer, etc.; and Cicero himself has left us a complete definition of the word, for he has described "sacellum" as "locus parvus deo sacratus cum ara."[1]

Again, in favour of the view that the existing building on Inchcolm is the actual chapel or oratory in which the insular anchorite lived and worshipped there in the twelfth century, it may be further argued, that, where they were not constructed of perishable materials, it was in consonance with the practice of these early times to preserve carefully houses and buildings of religious note, as hallowed relics. Most of the old oratories and houses raised by the early Irish and Scottish saints were undoubtedly built of wattles, wood, or clay, and other perishable materials, and of necessity were soon lost.[2] But when of a more solid and permanent

[1] *De Divinitate*, cap. 46.

[2] Though Roman houses, temples, and other buildings of stone and lime abounded in this country in the earlier centuries of the Christian era, yet the first Christian churches erected at Glastonbury in England, and at St. David's in Wales, were—according to the authority, at least, of William of Malmesbury and Giraldus Cambrensis—made of wattles. The first Christian church which is recorded as having been erected in Scotland, namely, the *Candida Casa*, reared at Whithern, towards the beginning of the fifth century, by St. Ninian, was constructed, as mentioned in a well-known passage of Bede, of stone, forming "ecclesiam insignem . . . . de lapide insolito Britonibus more."—(*Historia*

construction, they were sometimes sedulously preserved, and piously and punctually visited for long centuries as holy shrines. There still exist in Ireland various stone oratories of early Irish saints to which this remark applies—as, for example, that of St. Kevin at Glendalough, of St. Columba at Kells, those of St. Molua and St. Flannan at Killaloe, of St. Benan on Aranmore, St. Ceannanach on Inishmaan, etc. etc. Let us take the first two examples which I have named, to illustrate more fully my remark. St. Kevin died at an extreme old age, in the year 618; and St. Columba died a

Ecclesiast., lib. iii. cap. 4.) According to the *Irish Annals*, the three churches first erected by Palladius, in Ireland, about the year 420, were of wood, one of them being termed House of the Romans, "Teach-na-Romhan," but not apparently from its Roman mode of building.—(See Dr. O'Donovan's *Annals of the Four Masters*, vol. i. p. 129.) The church of Duleek, one of the earliest, if not the earliest, which St. Patrick erected in Ireland, and the first bishop of which, St. Cianan, died in the year 490, was built of stone, as its original name of Daimhllag (stone house) signifies; and the same word, *damhliag* or *stone house*, came subsequently to be applied as a generic term to the larger Irish churches.—(See Dr. Petrie's *Ecclesiastical Architecture of Ireland*, p. 142, with a quotation from an old Irish poem of the names of the three masons in the household of St. Patrick, who "made damhliags first in Erin.") When, in the year 652, Finan succeeded to the Bishopric of Lindisfarne, he built there a suitable Episcopal church, constructed of oak planks, and covered with reeds, "more Scotorum non de lapide, sed de robore secto totam composuit, atque arundine texit."—(Bede's *Hist. Eccl.*, lib. iii. cap. 25.) When St. Cuthbert erected his anchorite retreat on the island of Farne he made it of two chambers, one an oratory, and the other for domestic purposes; and he finished the walls of these buildings by digging round and cutting away the natural soil within and without, forming the roof out of rough wood and straw, "de lignis informibus et fœno.—(Vita S. Cuthberti, cap. 17.) Planks or "tabulæ," also, were employed in building or reconstructing the walls of this oratory on Farne Island, as St. Ethelwald, Cuthbert's successor, finding hay and clay insufficient to fill up the openings that age made be-

few years earlier, namely in the year 597. When speaking of the two houses at Glendalough and Kells, respectively bearing the names of these two early Irish saints, Dr. Petrie—and I certainly could not quote either a higher or a more cautious antiquarian authority—observes, "I think we have every reason to believe that the buildings called St. Columba's House at Kells, and St. Kevin's House at Glendalough, buildings so closely resembling each other in every respect, were erected by the persons whose names they

tween its boards, obtained a calf's skin, and nailed it as a protection against the storms in that corner of the oratory, where, like his predecessor, he used to kneel or stand when praying.—(*Ibid.*, cap. 46.) St. Godric's first rude hermitage at Finchale, on the Wear, was made of turf (vili cespite), and afterwards of rough wood and twigs (de lignis informibus et virgulis).—(See chaps 21 and 29 of his Life by Reginald.) On the construction, by wattles and wood, of some early Irish and Scoto-Irish monastic and saints' houses and oratories, as those of St. Wolloc, St. Columba, and St. Kevin, see Dr. Reeves' notes in his edition of the *Life of St. Columba*, pp. 106, 114, and 177. In some districts where wood was scarce, and stone abundant and easily worked, as in the west coast of Ireland, all ecclesiastical buildings were—like the far more ancient duns and forts in these parts—made principally or entirely of stone. But even in parts where wood was easily procured, oratories seem to have been sometimes, from an early period, built of stone. Thus, in the Tripartite Life of St. Patrick, the devout virgin Crumtherim is described as living in a stone-built oratory, "in cella sive *lapideo* inclusorio," in the vicinity of Armagh, as early as the fifth century.—(Colgan's *Trias Thaumaturga*, p. 163.) And, at the city of Armagh again, we have an incidental notice of a stone oratory in the eighth century; for, in the *Ulster Annals*, under the year 788, there is reported "Contentio in Ardmacae in qua jugulatur vir in hostio [ostio] Oratorii *lapidei*."—(Dr. O'Conor's *Rerum Hibernicarum Scriptores*, tom. iv. p. 113.) Dr. Petrie believes that all the churches at Armagh erected by St. Patrick and his immediate successors were built of stone, as well indeed as all the early abbey and cathedral churches throughout Ireland.—(*Ecclesiastical Architecture*, p. 159.)

bear."[1] If Dr. Petrie's idea be correct, and he repeats it elsewhere,[2] then these houses were constructed about the end of the sixth century, and their preservation for so long an intervening period was

[1] *The Ecclesiastical Architecture of Ireland*, anterior to the Anglo-Saxon Invasion, comprising an Essay on the Round Towers of Ireland, pp. 437, 435 and 430.

[2] "That these buildings (St. Columb's House at Kells and St. Kevin's at Glendalough), which are so similar in most respects to each other, are of a very early antiquity, can scarcely admit of doubt; indeed, I see no reason to question their being of the times of the celebrated ecclesiastics whose names they bear."—(Dr. Petrie's *Ecclesiastical Architecture*, p. 430.) In his late edition of Adamnan's *Life of St. Columba*, Dr. Reeves, when describing the Columbite monasteries and churches founded in Ireland, speaks (p. 278) of Kells as "having become the chief seat of the Columbian monks" shortly after the commencement of the ninth century. Among the indications of the ancient importance of the place which still remain, he enumerates the fine old Round Tower of Kells, its three ancient large sculptured crosses, the "curious oratory called St. Columbkille's House," and its great literary monument now preserved in Trinity College, Dublin—namely, the *Book of Kells*. He quotes the old Irish *Life of St. Columba*, followed by O'Donnell, to show that it is there stated that the saint himself "marked out the city of Kells in extent as it now is, and blessed it;" but he doubts if any considerable church here was founded by Columba himself, or indeed before 804. He grounds his doubts chiefly on the negative circumstance that there is "no mention of the place in the *Annals* as a religious seat" till the year 804. But the *Annals of the Four Masters* record two years previously, or in 802, that "the church of Columcille at Céanannus (or Kells) was destroyed" (vol. i. p. 413), referring of course to an *old* or former church of St. Columba's there; whilst the *Annals of Clonmacnoise* mention that two years afterwards, or in 804, "there was a *new* church founded in Kells in honour of St. Colume."—(See *Ibid.*, footnote.)* The learned editor of the *Annals of the Four Masters*, Professor O'Donovan, has translated and published, in the first volume of the *Miscellany of the Irish Archæological Society*, an ancient poem attributed to St. Columba, and

* [St. Colume, as translated by Mageochagan or Macgeoghegan. In the original this would be Columbkille, as in all the other Annals.—P.]

no doubt in a great measure the result of their being looked upon, protected, and visited, as spots hallowed by having been the earthly dwellings of such esteemed saints.

In the great work on *The Ecclesiastical Architecture of Ireland,*

which, at all events, was certainly composed at a period when some remains of Paganism existed in Ireland. In this production the poet makes St. Columba say, "My order is at Cennanus (Kells)," etc.; and in his note to this allusion Dr. O'Donovan states that at Kells "St. Columbkille erected a monastery in the sixth century."—(*Miscellany of Archæological Society*, vol. i. p. 13.) Some minds would trust such a question regarding the antiquity of a place more to the evidence of parchment than to the evidence of stone and lime. The beautiful *Evangeliarium,* known as the *Book of Kells,* is mentioned by the *Four Masters* under the year 1006 as being then the "principal relic of the western world," on account of its golden case or cover, and as having been temporarily stolen in that year from the erdomh or sacristy of the great church of Kells. In the same ancient entry this book is spoken of as "the Great Gospel of Columcille," and whether originally belonging to Kells or not, is certainly older than the ninth century, if not indeed as old as Columba. The corresponding *Evangeliarium* of Durrow, placed now also in Trinity College, Dublin,—"a manuscript" (says Dr. Reeves, p. 276) "approaching, if not reaching to the Columbian age,"—is known from the inscription on the silver-mounted case which formerly belonged to it, to have been "venerable in age, and a reliquary in 916" (p. 327). In the remarkable colophon which closes this manuscript copy of the Evangelists, St. Columba himself is professed to be the copyist or writer of it, the reader being adjured to cherish the memory "Columbæ scriptoris *qui hoc scripsi.*" In the *Ulster Annals,* under the year 904, there is the following entry regarding Kells: "Violatio Ecclesiæ Kellensis per Flannum mac Maelsechnalli contra Donchad filium suum, et alii decollati sunt circa *Oratorium.*"—(Dr. O'Conor's *Rerum Hibern. Scriptores,* tom. iv. p. 243.) Is the scene of slaughter thus specialised the Oratory or "House of St. Columb," which is still standing at Kells? *

---

* [I would say yes, beyond question! It was both oratory and house, like that of St. Cuthbert on Farne island, described in the passage quoted *ante*, p. 101, note.—P.]

which I have just quoted—a work, let me add, overflowing with the richest and ripest antiquarian lore, and yet written with all the fascination of a romance—Dr. Petrie, after describing the two houses I speak of, St. Kevin's and St. Columba's, farther states his belief that both of these buildings "served the double purpose of a habitation and an oratory."[1] They were, in this view, the residences, as well as the chapels, of their original inhabitants; and subsequently the house of St. Kevin at Glendalough, of St. Flannan at Killaloe, etc., were publicly used as chapels or churches.[2] In all probability the *capellula* of the hermit on Inchcolm was, in the same way, at once both the habitation and the oratory of this solitary anchorite, and apparently the only building on the island when Alexander was tossed upon its shores. The sacred character of the humble cell, as the dwelling and oratory of a holy Columbite hermit, and possibly also the interest attached to it as an edifice which had afforded for three days such welcome and grateful shelter to King Alexander and his suite, would in all probability—judging from the numerous analogies which we might trace elsewhere—lead to its preservation, and perhaps its repair and restora-

[1] In treating of the subsequent fate of the old Irish oratories, Dr. Petrie remarks, "Such structures came in subsequent times to be used by devotees as penitentiaries, and to be generally regarded as such exclusively. Nor is it easy to conceive localities as such better fitted, in a religious age, to excite feelings of contrition for past sins, and of expectations of forgiveness, than those which had been rendered sacred by the sanctity of those to whom they had owed their origin. Most certain, at all events, it is, that they came to be regarded as sanctuaries the most inviolable, to which, as our annals show, the people were accustomed to fly in the hope of safety—a hope, however, which was not always realised."—(P. 358.)

[2] *Scotichronicon*, lib. v. cap. 36. Goodall's edition, vol. i. p. 286.

tion, when, a few years afterwards, the monastery rose in its immediate neighbourhood, in pious fulfilment of the royal vow.*

Indeed, that the holy cell or chapel of the Inchcolm anchorite would, under the circumstances in question, be carefully saved and preserved by King Alexander I., is a step which we would specially expect, from all that we know of the religious character of that prince, and his peculiar love for sacred buildings and the relics of saints. For, according to Fordun, Alexander "vir literatus et pius" "erat in construendis ecclesiis, et reliquis Sanctorum perquirendis, in vestibus sacerdotalibus librisque sacris conficiendis et ordinandis, studiosissimus."

For the antiquity of the Inchcolm cell there yet remains an additional argument, and perhaps the strongest of all. I have already

* [Such cells or oratories, as relics of the holy men who had been their founders, were always regarded by the Irish, like every other kind of relics, as their bells, croziers, books, etc. etc., with the deepest sentiments of veneration, and their injury or violation—"dishonouring," as the annalists often term it—was regarded as a sacrilege of the most revolting and sinful character. And to this pious feeling we may ascribe the singular preservation to our own times of so many of such buildings—though, indeed, in many instances, they may only retain the general form, or a portion of the walls, of the original structure—owing to the injuries inflicted by time, or, as more frequently, by foreign violence. Thus, in the great Aran of the *Tiglach Enda*, or "House of Enda," a portion only—the east end—is of the Saint's time, the rest is some centuries later ; and of St. Ciarn's oratory at Clonmacnoise—called in the *Irish Annals* "Temple Ciaron," or "Eaglais-beag," and, sometimes, "*Temple-beg*," or "The Little Church," though the original form was carefully preserved, there was, when I first examined it, more than forty years ago, apparently no portion of its masonry that was not obviously of much later times—in parts even as late as the seventeenth century. Our annalists record the names of Airchinneachs of this oratory from 893 to 1097.—P.]

stated that, in its whole architectural type and features, the cell or oratory is manifestly older, and more rude and primitive, than any of the diverse monastic buildings erected on the island from the twelfth century downwards. But more, the Inchcolm cell or oratory corresponds in all its leading architectural features and specialities with the cells, oratories, or small chapels, raised from the sixth and eighth, down to the tenth and twelfth centuries, in different parts of Ireland, and in some districts in Scotland, by the early Irish ecclesiastics, and their Irish or Scoto-Irish disciples and followers, of these distant times and dates.

It is now acknowledged on all sides, that, though not the first preachers of Christianity in Scotland,[1] the Irish were at least by

[1] In reference to this observation, it is scarcely necessary to refer to the teachings in Scotland of St. Kentigern of Strathclyde in the first half of the sixth century, of St. Serf of Culross in the latter, and of St. Palladius and St. Ninian in the earlier parts of the fifth century, with the more immediate converts and followers of these ancient missionaries. In his *Demonstratio quod Christus sit Deus contra Judæos atque Gentiles*, written about the year 387, St. Chrysostom avers that "the British Islands (Βρεττανικαὶ νῆσοι), situated beyond the Mediterranean Sea, and in the very ocean itself, had felt the power of the Divine Word, churches having been found there, and altars erected." (*Opera omnia*, vol. i. p. 575, Paris edition of Montfaucon, 1718.) Perhaps St. Chrysostom founded his statement upon a notice in reference to the alleged extension of Christianity to the northern parts of Britain, given a hundred and fifty years previously by Tertullian, when discussing a similar argument. In his dissertation *Adversus Judæos*, supposed to be written about 210, Tertullian, when treating of the propagation of Christianity, states (chap. vii.), that at that time already places among the Britons, inaccessible to the Romans, were yet subject to Christ—"Britannorum inaccessa Romanis loca, Christo vero subdita." (Oehler's edition of *Tertullian*, vol. iii. p. 713.) Among the numerous inscriptions and sculptures left here by the Romans while they held this country during the first four centuries of the Christian

far the most active and the most influential of our early mission aries; and truly a new epoch began in Scottish history when, in the year 563, St. Columba, "pro Christo peregrinari volens," embarked, with his twelve companions, and sailing across from Ireland to the west coast of Scotland, founded the monastery of Iona. It is certainly to St. Columba and his numerous disciples and followers that the spread of Christianity in this country, during the succeeding two or three centuries, is principally due. At the same time we must not forget that numerous other Irish saints in these early times engaged in missionary visits to Scotland, and founded churches there, which still bear their names, as (to quote part of the enumeration of Dr. Reeves) St. Finbar, St. Comgall, St. Blaan, St. Brendan, the two St. Fillans, St. Ronan, St. Flannan,

era, not one has, I believe, been yet found containing a single Christian notice or emblem, or affording by itself any direct evidence of the existence of Christianity among the Roman colonists and soldiers in Britain. But there is indirect lapidary or monumental evidence of its propagation in another manner. In England, as in Germany, France, etc., there exist among the old Roman remains, altars and temples dedicated to Mithras, originally the god of the Sun among the Persians, with sculptures and inscriptious referring to Mithraic worship. They have been found in the cities along the Roman wall in Northumberland; at York, etc. Various references among the old Fathers seem to show that when a knowledge of the Christian religion began to spread to the Western Colonies of Rome, the worship of Mithras was set up in opposition to Christianity, and Christian rites were imitated by the Mithraic priests and followers. Thus, for example, the author whom I have just cited, Tertullian, tells us, in his tract *De Præscriptione Hæreticorem*, chap. 40, that the worshippers of Mithras practised the remission of sins by water (as in baptism), made a sign upon their foreheads (as if simulating the sign of the cross), celebrated the offering of bread (as if in imitation of the sacrament of the Lord's Supper), etc. (See his *Works*, vol. iii. p. 38, of Oehler's Leipsic edition of 1854.)

St. Beranch, St. Catan, St. Merinus, St. Mernoc, St. Molaise, St. Munna, St. Vigean, etc.[1]

Along with their Christian doctrines and teachings these Irish ecclesiastics brought over to Scotland their peculiar religious habits and customs, and, amongst other things, imported into this country their architectural knowledge and practices with regard to sacred and monastic buildings. In the western parts of Scotland, more particularly, numerous ecclesiastical structures were raised similar to those which were peculiar to Ireland; and various material vestiges of these still exist.* In the eastern parts of Scotland, to which the personal teaching of the Irish missionaries speedily spread, we have still remaining two undoubted examples of the repetition in this country of Irish ecclesiastical architecture in the well-known Round Towers of Abernethy and Brechin, and perhaps we have a third example in the stone-roofed oratory of Inchcolm.

Various ancient stone oratories still exist in a more or less perfect condition in different parts of Ireland, sometimes standing by themselves, sometimes with the remains of a round beehive-shaped cell or dwelling near them, and sometimes forming one of a group of churches, or of a series of monastic buildings. Such, for example, are the small chapels or oratories of St. Gobnet, St. Benen,

[1] See Dr. Reeves' admirable edition of Adamnan's *Life of St. Columba*, pp. lxxiv and lxxv,—a book which is a perfect model of learned annotation and careful editing.

---

* [I think it might be well to strengthen your statement by adducing a few examples—thus, as for example, the remains of a monastery of Columba's time on an island—now drained—called Lough Columbkill, in the island of Skye—the churches and clochans, or stone-houses of the monks, on St. Kilda, and probably many similar remains on other islands of the Hebrides.—P.]

and St. MacDuach, in the Isles of Aran,* of St. Senan on Bishop's Island, of St. Molua on Friar's Island, Killaloe, the Leabha Mollayga near Mitchelstown, in the County Cork, and probably the so-called dormitory of St. Declan at Ardmore. Among the old sacred buildings of Ireland we find, in fact, two kinds or classes of churches, the "ecclesiæ majores" and "minores," if we may call them so, and principally distinguished from each other by their comparative length or size. It appears both from the remains of the first class which still exist, and from the incidental notices which occur of their erection, measurements, etc., in the ancient annals and hagiology of Ireland, that the larger abbey or cathedral churches of that country, whose date of foundation is anterior to the twelfth century, were oblong quadrangular buildings, which rarely, if ever, exceeded the length of 60 feet, and were sometimes less. In the Tripartite Life of St. Patrick, he is described as prescribing 60 feet as the length of the church of Donagh Patrick.[1] This "was also," says Dr. Petrie, "the measure of the other celebrated chapels erected by him throughout Ireland, and imitated as a model by his successors."[2] "Indeed," he further observes, "that the Irish, who have been ever remarkable for a tenacious adherence to their ancient customs, should preserve with religious veneration that form and size of the primitive church introduced by the first teachers of Christianity, is only what might be naturally expected, and what we find to have been the fact. We

[1] Colgan's *Trias Thaumaturga*, p. 129.

[2] *Ecclesiastical Architecture of Ireland*, p. 195.

---

* [Of St. MacDara of Cruach MicDara, an island off the coast of Connamara, of St. Brendan in Inis Gloria, an island off the coast of Errus, and very many more.—P.]

see," Dr. Petrie adds, "the result of this feeling exhibited very remarkably in the conservation, down to a late period, of the humblest and rudest *oratories* of the first ecclesiastics in all those localities where Irish manners and customs remained, and where such edifices, too small for the services of religion, would not have been deemed worthy of conservation, but from such feeling."[1]

The second or lesser type of the early Irish churches, or, in other words, of the humble and rude oratories to which Dr. Petrie refers in the last sentence of the preceding paragraph, were of a similar form, but of a much smaller size than the larger or abbey churches.* We have ample and accurate evidence of this, both in the oratories which still remain, and in a fragment of the Brehon laws, referring to the different payments which ecclesiastical artificers received according as the building was—(1.) a duirtheach or small chapel or oratory ; (2.) a large abbey church or damhliag, etc.[2]

Generally, according to Dr. Petrie, the average of the smaller type of churches or oratories may be stated to be about 15 feet in length, and 10 feet in breadth, though they show no fixed similarity in regard to size.[3] "In the general plan," he observes, "of this class of buildings there was an equal uniformity. They had a single doorway, always placed in the centre of the west wall,† and were

[1] *Ecclesiastical Architecture of Ireland*, p. 194.
[2] *Ibid.*, pp. 365, 351. [3] *Ibid.*, p. 351.

* [And which, moreover, had often chancels attached to them.—P.]

† [I should, perhaps, have written *almost* always. The very few exceptions did not at the moment occur to me. Perhaps, indeed, there is but one exception, that most important one, on Bishop's Island, the others belonging rather to churches.—P.]

lighted by a single window placed in the centre of the east wall, and a stone altar usually, perhaps always, placed beneath this window."[1] In these leading architectural features (with an exception to which I shall immediately advert), the Inchcolm cell or oratory corresponds to the ancient cells or oratories existing in Ireland, and presents the same ancient style of masonry—the same splaying internally of the window which is so common in the ancient Irish churches, both large and small—and the same configuration of doorway which is seen in many of them, the opening forming it being narrower at the top than at the bottom.

Fig. 10. St. Senan's Oratory on Bishop's Island.

In the Inchcolm oratory there is one exception, as I have just stated, to the general type and features of the ancient Irish oratory. I allude to the position of the door, which is placed in the south side of the Inchcolm cell, instead of being placed, as usual, in the western gable of the building. But this position of the door in the south wall is not without example in ancient Irish oratories that still exist.* The door occupies in this respect the same position in

[1] *Ecclesiastical Architecture of Ireland*, p. 352.

* [South doorways are certainly very rarely to be met with in the very ancient churches or oratories in Ireland. In addition to this important one

the Inchcolm oratory as in an oratory on Bishop's Island upon the coast of Clare, the erection of which is traditionally ascribed to St. Senan, who lived in the sixth century. This oratory of St. Senan (says Mr. Wakeman) "measures 18 feet by 12; the walls are in thickness 2 feet 7 inches. The doorway, which occupies an unusual position in the south side, immediately adjoining the west end wall, is 6 feet in height, and 1 foot 10 inches wide at the top, 2 feet 4 inches at the bottom. The east window splays externally, and in this respect is probably unique in Ireland."[1] * These peculiarities are shown in the accompanying woodcut, Fig. 10, taken from Mr. Wakeman's *Handbook of Irish Antiquities.*

The Irish ecclesiastics did not scruple to deviate from the established plans of their sacred buildings, when the necessities of individual cases required it. In the Firth of Forth west winds are the most prevalent of all; and sometimes the western blast is still as fierce and long continued as when of old it drove King Alexander on the shores of Inchcolm. The hermit's cell or oratory is placed on perhaps the most protected spot on the island; and yet it would

[1] Wakeman's *Archæologia Hibernica*, pp. 59, 60.

---

on Bishop's Island, I can only call to mind three others, namely, in Kilbaspugbrone, near Sligo; the Templemor, or great church of St. Mochonna, in Inismacnerin, or, as now called, Church Island, in Lough Key, county of Roscommon; and Killcrony, near Bray, in the county of Wicklow. The two last named are fine specimens of doorways of Cyclopean style and masonry.—P.]

* [My pupil is in error in this supposition. He should have remembered —for he drew it on the block for me—that the window in the oratory near the church of Kilmalkedar, county of Kerry, which is built without cement, splays both externally and internally.—See my work, p. 184.

I should also observe another feature common to both these windows, namely, that it is only the jambs that are splayed. —P.]

have been scarcely habitable with an open window exposing its interior to the east, and with a door placed directly opposite it in the western gable. It has been rendered, however, much more fit for a human abode by the door being situated in the south wall; and the more so, because the ledge of rock against which the south-west corner of the building abuts, protects in a great degree this south door from the direct effects of the western storm. The building itself is narrower than the generality of the Irish oratories, but this was perhaps necessitated by another circumstance, for its breadth was probably determined by the immovable basaltic blocks lying on either side of it.

The head of the doorway in the Inchcolm oratory is, as pointed out in a preceding page, peculiar in this respect, that externally it is constructed on the principle of the radiating arch, whilst internally it is built on the principle of the horizontal arch. But in other early Irish ecclesiastical buildings in Scotland, as well as in Ireland, the external and internal aspect of the doorway is sometimes thus constructed on opposite principles. In the round tower, for example, of Abernethy, the head of the doorway externally is formed of a large single stone laid horizontally, and having a semicircular opening cut out of the lower side of the horizontal block; while the head of the doorway internally is constructed of separate stones on the plan of the radiating arch.

One striking circumstance in the Inchcolm oratory—viz., its vaulted or arched roof, has been already sufficiently described; and, in describing it, I have stated that the arch is of a pointed form. In many of the ancient Irish oratories the roof was of wood, and covered with rushes or shingles; and most of them had their walls even constructed of wood or oak, as the term *duir-theach* originally signifies. But apparently, though the generic name duir-theach still continued to be applied to them, some of them were constructed, from a very early period, entirely of stone; and of these the roofs were occasionally formed of the same material as the walls, and arched or vaulted, as in the Inchcolm oratory. In speaking of the construction of the primitive larger churches of Ireland, Dr. Petrie states, that their "roof appears to have been constructed generally of wood, even where their walls were of stone;" while in the oratories or primitive smaller stone churches, "the roofs (says he) generally appear to have been constructed of stone, their sides forming at the ridge a very acute angle."[1] The selection of the special materials of which both walls and roof were composed, was no doubt, in many cases, regulated and determined by the comparative facility or difficulty with which these materials were obtained. At no time, perhaps, did timber exist on Inchcolm that could have been used in constructing such a building; whilst plenty of stones fit for the purpose abounded on the island, and there was abundance of lime on the neighbouring shore. Stone-roofed oratories of a more complex and elaborate architectural character than that of Inchcolm still exist in Ireland, and of a supposed very early date. We have already found, for instance, Dr. Petrie stating that "we have every reason to believe" that the stone-roofed oratories known as St. Kevin's House at Glendalough, and St. Columba's House at Kells, "were erected by the persons whose names they bear,"[2] and consequently that they are as old as the sixth century. These two oratories, are, as it were, two storeyed buildings; for each consists of

[1] *Ecclesiastical Architecture of Ireland*, p. 186. [2] *Ibid.* p. 437.

a lower and larger stone-arched or vaulted chamber below, and of another higher and smaller stone-arched or vaulted chamber or over-croft above. The old small stone-roofed church still standing at Killaloe, and the erection of which Dr. Petrie is* inclined to ascribe to St. Flannan in the seventh century† presents also in its structure this type of double stone-vault or arch, as shown in the following section of it by Mr. Fergusson.[1] When treating of the early Irish

Fig. 11. Section of St. Flannan's Church at Killaloe.

oratories, Mr. Fergusson observes, "One of the peculiarities of these churches is, that they were nearly all designed to have stone roofs, no wood being used in their construction. The section (Fig. 11) of the old church at Killaloe, belonging probably to the tenth century, will explain how this was generally managed. The nave was roofed

[1] See his *Illustrated Handbook of Architecture*, vol. ii. p. 918.

* [Was.—P.] † [But now considers as of the tenth or perhaps eleventh.—P.]

with a tunnel-vault with a pointed one over it, on which the roofing slabs were laid." Mr. Fergusson adduces Cormac's Chapel on the Rock of Cashel, St. Kevin's House or Kitchen at Glendalough, which he thinks "may belong to the seventh century;" and St. Columba's House at Kells, "and several others in various parts of Ireland, as all displaying the same peculiarity" in the stone roofing.

Like some oratories and churches in Ireland, more simple and primitive than those just alluded to, the building on Inchcolm is an edifice consisting of a single vaulted chamber, analogous in form to the over-croft of the larger oratories or minor churches. The accompanying section of the old and small stone-roofed church of

Fig. 12. Section of the Church of Killaghy.

Killaghy, at the village of Clonghereen, near Killarney, is the result of an accurate examination of that building by Mr. Brash of Cork. Its stones look better dressed and more equal in size, but otherwise it is so exactly a section of the Inchcolm oratory, that it might well be regarded as a plan of it, intended to display the figure and mode of construction of its walls and stone roof, formed as that roof is of three layers—viz., 1. The layer consisting of the proper stones of the arch of the cell interiorly; 2. The layer of outer roofing stones placed exteriorly; and 3. The intermediate layer of lime, and grit or small stones, cementing and binding together these other two courses.*

It was once suggested to me as an argument against the Irish architectural character and antiquity of the Inchcolm oratory, that its vault or arch was slightly but distinctly pointed, and that pointed arches did not become an architectural feature in ecclesiastical buildings before the latter half of the twelfth century. But if there existed any truth in this objection, it would equally disprove the early character and antiquity of those ecclesiastical buildings at Killaloe, Glendalough, and Kells, in which the arch of the over-croft is of the same pointed form. The over-croft in King Cormac's Chapel at Cashel shows also a similar pointed vault or arch; and no one now ventures to challenge it as an established fact in ecclesiological history, that this edifice was consecrated in 1134, or at a date anterior to the introduction[1] of Gothic church architecture or pointed arches in sacred buildings in England.[2] In truth, the

[1] See Dr. Petrie's work (p. 291) for full quotations in confirmation of this date, from the *Annals of Clonmacnoise and Kilronan*, the *Annals of Munster*, the *Annals of the Four Masters*, the *Chronicon Scotorum*, etc.

[2] When discussing the history of the pointed arch, Mr. Parker observes: "The choir of Canterbury Cathedral, commenced in 1175, is usually referred to as the earliest example in England, and none of earlier date has been authenticated."—*Glossary of Terms in Architecture* (1845), p. 28.

---

* [I confess that I should not like to adduce this stone-roofed church of Killaghy in support of the antiquity of the oratory; for I could never bring myself to believe that it was of an age anterior to the thirteenth century.—P.]

pointed form of arched vault was sometimes used by Irish ecclesiastics structurally, and for the sake of more simply and easily sustaining the stone roof, long before that arch became the distinctive mark of any architectural style. Indeed, in the very oldest existing Irish oratory—viz. that of Gallerus, which is generally reckoned[1] as early as, if not earlier than, the time of St. Patrick, or about the fifth century—the stone roof, though constructed on the principle of the horizontal arch, is of the pointed form. The whole section of the oratory of Gallerus is that of a pointed arch commencing directly at the ground line.[2] "I have," Mr. Brash writes me, and I could not well quote a better judge or more learned ecclesiastic antiquary, "carefully examined the oratory at Inchcolm, and it is my conviction that the pointed arch supporting the stone roof does not in any wise whatever militate against its antiquity, particularly when taking it in connection with the extreme rudeness and simplicity of the rest of the structure, and the total absence of any pointed form in either door or window."*

Let me add one word more as to the probable or possible age of the capellula on Inchcolm. Granting, for a moment, that the building on Inchcolm is the small chapel existing on the island when visited by King Alexander in 1123, have we any reason to

[1] Dr. Petrie's *Ecclesiastical Architecture*, p. 133.

[2] Pointed arches, constructed both on the radiating and horizontal principles, are found still standing in the antiquated mason-work of Assyria, Nubia, Greece, and Etruria. (See drawings and descriptions of different specimens from these countries in Mr. Fergusson's *Handbook of Architecture*, vol. i. pp. 253, 254, 257, 259, 294, 381, etc.) The pointed arch was used in the East in sacred architecture as early as the time of Constantine, as is

* [In this opinion of Mr. Brash's I fully concur.—P.]

suppose the structure to be one of a still earlier date? Inchcolm was apparently a favourite place of sepulture up, indeed, to comparatively late times; and may possibly have been so in old Pagan times, and previously to the introduction of Christianity into Scotland. The soil of the fields to the west of the monastery is, when turned over, found still full of fragments of human bones. Allan de Mortimer, Lord of Aberdour, gave to the Abbey of Inchcolm a moiety of the lands of his town of Aberdour for leave of burial in the church of the monastery.[1] In Scottish history various allusions

still witnessed in the oldest existing Christian church, namely, the church built by that emperor, in the earlier part of the fourth century, over the alleged tomb of our Saviour at Jerusalem.* For notices of the prevalence of the pointed arch in early Eastern and in Saracenic architecture, see Fergusson's *Handbook*, p. 380, 598, etc.

[1] "Alanus de Mortuo Mari, Miles, Dominus de Abirdaur, dedit omnes et totas dimidietates terrarum Villæ suæ de Abirdaur, Deo et Monachis de Insula Sancti Columbi, pro sepultura sibi et posteris suis in Ecclesia dicti Monasterii." (Quoted from the MS. Register or Chartulary of the Abbey by Sir Robert Sibbald in his *History of Fife*, p. 41.) The same author adds, that, in consequence of this grant to the Monastery of Inchcolm for leave of sepulture, the Earl of Murray (who represents "Stewart Abbott of Inchcolm," that sat as a lay Commendator in the Parliament of 1560, when the Confession of Faith was approved of) now possesses "the wester half of Aberdour." Sir Robert Sibbald further mentions the story that "Alain, the founder, being dead, the Monks, carrying his corpse in a coffin of lead, by barge, in the night-time, to be interred within their church, some wicked Monks did throw the samen in a great deep betwixt the land and the Monastery, which to this day, by the

* [I must confess that I am very sceptical as to any portion now existing of the church of the Holy Sepulchre being of the time of Constantine, and also as to the early age of any portion of it in which a pointed arch is found. More walls of the original edifice may *possibly* exist; but it is certain that the church was more than once modified, and the ornamental work is assuredly of a much later age.—P.]

occur with regard to persons of note, and especially the ecclesiastics of Dunkeld, being carried for sepulture to Inchcolm.[1] The Danish chiefs who, after the invasion of Fife, were buried in the cemetery of Inchcolm, were, as we have already found, interred there in the seventh or last year of King Duncan's reign, or in 1039, nearly a century before the date of Alexander's visit to the island. But if there was, a century before Alexander's visit, a place of burial on the island, there was almost certainly also this or some other chapel attached to the place, as a Christian cemetery had in these early times always a Christian chapel or church of some form attached to it. The style and architecture of the building is apparently, as I have already stated, as old, or even older than this; or,

neighbouring fishermen and salters, is called *Mortimer's deep.*" He does not give the year of the preceding grant by Alain de Mortimer, but states that "the Mortimers had this Lordship by the marriage of Anicea, only daughter and sole heiress of Dominus Joannes de Vetere Ponte or Vypont, in anno 1126." It appears to have been her husband who made the above grant. (See Nisbet's *Heraldry*, vol. i. p. 294.)

[1] Thus, in 1272, Richard of Inverkeithing, Chamberlain of Scotland, died, and his body was buried at Dunkeld, but his heart was deposited in the choir of the Abbey of Inchcolm. (*Scotichronicon*, lib. x. c. 30.) In Hay's *Scotia Sacra* is a description of the sepultures on this monument in Inchcolm Church, p. 471. In 1173, Richard, chaplain to King William, died at Cramond, and was buried in Inchcolm. (Mylne's *Vitæ*, p. 6.) In 1210, Richard, Bishop of Dunkeld, died at Cramond, and was buried in Inchcolm. (*Scotichronicon*, lib. viii. c. 27); and four years afterwards, Bishop Leycester died also at Cramond, and was buried at Inchcolm (*Ibid.* lib. ix. c. 27). In 1265, Richard, Bishop of Dunkeld, built a new choir in the church of St. Columba on Inchcolm; and in the following year the bones of three former bishops of Dunkeld were transferred and buried, two on the north, and the third on the south side of the altar in this new choir. (*Scotichronicon*, lib. x. c. 20, 21.) See also the *Extracta e Cronicis Scocie* for other similar notices, pp. 90, 95, etc.; and Mylne's *Vitæ Dunkeldensis Ecclesiæ Episcoporum*, pp. 6, 9, 11, etc.

at all events, it corresponds in* its features to Irish houses and oratories that are regarded as having been built two or three centuries before the date even of the sepulture of the Danes in the island.

The manuscript copy of the *Scotichronicon*, which belonged to the Abbey of Cupar, and which, like the other old manuscripts of the *Scotichronicon*, was written before the end of the fifteenth century,[1] describes Inchcolm as the temporary abode of St. Columba himself,† when he was engaged as a missionary among the Scots and

[1] "There are" (observes Father Innes) "still remaining many copies of Fordun, with continuations of his history done by different hands. The chief authors were Walter Bower or Bowmaker, Abbot of Inchcolm, Patrick Russell, a Carthusian monk of Perth, *the Chronicle of Cupar*, the Continuation of Fordun, attributed to Bishop Elphinstone, in the Bodleian Library, and many others. All these were written in the fifteenth age, or in the time betwixt Fordun and Boece, by the best historians that Scotland then afforded, and unquestionably well qualified for searching into, and finding out, what remained of ancient MSS. histories anywhere hidden within the kingdom, and especially in abbeys and monasteries, they being all either abbots or the most learned churchmen or monks in their respective churches or monasteries." (Innes' *Critical Inquiry*, vol. i. p. 228.)

---

* [Many, if not all of.—P.]

† [I confess I have still some doubt as to this island having received its name from a church founded by S. Columba-*cill*, or that he ever resided in it, and I should like to have your present opinion upon the matter. Fordun *alone* seems to me a very insufficient authority for a fact which is very improbable; and the legend of the seal, which I published, appears to me to be a better authority for the ancient name of the island—"*Colmanus nomine, qui ab alijs Mocholmocus.* Quia Colmôc & Colmân sunt diminutiva, a *Colum.* 1. Columba, et affectus vel venerationis causa additur *mo;* et hinc *Mocholmocus*," Colgan, vol. i. p. 155. Colgan's authority is of no value, as his statement is wholly founded on Fordun. This is proved by his notice of the monastery in his catalogue of the churches founded by Columba. "Colmis-inse Monasterium canonicorum Regularium in Æmonia insula inter Edinburgum et Inver-

Picts. In enumerating the islands of the Firth of Forth, Inchcolm is mentioned in the Cupar manuscript as "alia insuper insula ad

---

Kithin. *Fordonus, ibid.*" As the cautious Dr. Lanigan observes—"Colgan was, to use a vulgar phrase, bewitched as to the mania of ascribing foundations of monasteries to our eminent saints." Further, it should not be forgotten that Fordun tells us that in his time the island was called "*Saint Colmy's Inche.*" See the passage quoted by Ussher, *De Brit. Ec.*, p. 704. Now, I know of no instance of the corruption of Columb, or Columba, into Colmy, which appears rather a corruption of Colmoc or Colman.

If this be not the Insula Colmoci of the *regal* seal—"round seals have something royal,"—where are we to find it? Not in Ireland, certainly, though our calendars record the names of two islands called Inch Mocholmoc, from saints of that name. One of these was in Leinster; the locality of the other is unknown. They also record the patron day of a St. Mocholmoc, *na hainse*, "of the island," at the 30th October. Could we find what was the patron day of the saint of Inche Colm it might help to settle the matter. One of the above saints is called Colman *Ailither*, or the pilgrim. Chattering in my discursive way, let me add that a Saint Mocholmoc appears to have been a favourite with the Danes of Dublin in the twelfth century, for we find in the lists of the Danish Kings of Dublin that of Donald MacGilloholmoch as reigning from 1125 to 1134; and another of the name is noticed by Regan as an Irish king, who lived not far from Dublin, and who offered his services to the English against the Irish and Danes in 1171. There was a Gillmeholmoc's Lane in Dublin, near Christ's Church, where, as Harris conjectures, he, or some of his family, inhabited. Did this royal Danish family adopt its surname in honour of St. Colman of Lindisfarne, of whom it must have heard a great deal during the Danish occupation of Northumbria, the kings of which were for a long time also kings of Dublin? Or may it have been from a remembrance of the shelter and honourable interment to their dead, given to their predecessors in the little island of St. Colme (or Colmoch!) something more than a century before—said island having derived its name from the Lindisfarne Saint, who may have occasionally occupied it as his desert or hermitage? I do not expect that you will not laugh at all this! but a hearty laugh is not a bad thing in this gloomy weather.—P.]

occidens distans ab Inchcketh, quæ vocatur Æmonia, inter Edinburch et Inverkethyn; *quam quondam incoluit, dum Pictis et Scotis fidem prædicavit, Sanctus Columba Abbas.*"[1] We do not know upon what foundation, if any, this statement is based; but it is very evidently an allegation upon which no great assurance can be placed. Nor, in alluding to this statement here, have I any intention of arguing that this cell might even have served St. Columba both as a house and oratory, such as the house of the Saint still standing at Kells is believed by Dr. Petrie to have possibly been.

The nameless religious recluse whom Alexander found residing on Inchcolm is described by Fordun and Boece as leading there the life of a hermit (*Eremita*), though a follower of the order or rule of Saint Columba. The ecclesiastical writers of these early times not unfrequently refer to such self-denying and secluded anchorites. The Irish Annals are full of their obits. Thus, for example, under the single year 898, the Four Masters[2] record the death of, at least, four who had passed longer or shorter periods of their lives as hermits, namely, "Suairleach, anchorite and Bishop of Treoit;" "Cosgrach, who was called Truaghan [the meagre], anchorite of Inis-Cealtra;" "Tuathal, anchorite;" "Ceallach, anchorite and Bishop of Ard-Macha;"—and probably we have the obit of a fifth entered in this same year under the designation of "Caenchomhrac of the Caves of Inis-bo-finc," as these early ascetics sometimes betook themselves to caves, natural or artificial, using them for their houses

[1] See extract in Goodall's edition of the *Scotichronicon*, vol. i. p. 6. (foot-note), and in Colgan's *Trias Thaumaturga*, vol. ii. p. 466.

[2] Dr. O'Donovan's *Annals of the Kingdom of Ireland*, vol. i. p. 557.

and oratories.[1] Various early English authors also allude to the habitations and lives of different anchorites belonging to our own country. Thus the venerable Bede—living himself as a monk in the Northumbrian monastery of Jarrow, in the early part of the eighth century—refers by name to several, as to Hemgils, who, as a religious solitary (*solitarius*), passed the latter portion of his life sustained

[1] In Scotland we have various alleged instances of caves being thus employed as anchorite or devotional cells, and some of them still show rudely cut altars, crosses, etc.—as the so-called cave of St. Columba on the shores of Loch Killesport in North Knapdale, with an altar, a font or piscina, and a cross cut in the rock (*Origines Parochiales*, vol. ii. p. 40); the cave of St. Kieran on Loch Kilkerran in Cantyre (*Ibid.* vol. ii. p. 12); the cave of St. Ninian on the coast of Wigtonshire (*Old Statistical Account of Scotland*, vol. xvii. p. 594); the cave of St. Molio or Molaise, in Holy Island, in the Clyde, with Runic inscriptions on its walls (see an account of them in Dr. Daniel Wilson's admirable *Prehistoric Annals of Scotland*, pp. 531 to 533, etc). The island of Inchcolm pertains to Fifeshire, and in this single county there are at least four caves that are averred to have been the retreats which early Christian devotees and ascetics occupied as temporary abodes and oratories, or in which they occasionally kept their holy vigils; namely, the cave at Dunfermline, which bears the name of Malcolm Canmore's devout Saxon queen St. Margaret, and which is said to have contained formerly a stone table or altar, with "something like a crucifix" upon it (Dr. Chalmers' *Historical Account of Dunfermline*, vol. i. pp. 88, 89); the cave of St. Serf at Dysart (the name itself—Dysart—an instance, in all probability, of the "*desertum*" of the text, p. 124), in which that saint contested successfully in debate, according to the *Aberdeen Breviary*, with the devil, and expelled him from the spot (see *Breviarium Aberdonense*, Mens. Julii, fol. xv, and Mr. Muir's *Notices of Dysart* printed for the Maitland Club, p. 3); the caves of Caplawchy, on the east Fifeshire coast, marked interiorly with rude crosses, etc., and which, according to Wynton, were inhabited for a time by "St. Adrian wyth hys cumpany" of disciples (*Orygynale Chronykel of Scotland*, book iii. c. viii.); and the cave of St. Rule at St. Andrews, containing a stone table or altar on its east side, and on its west side the supposed sleeping cell of the hermit excavated out of the rock (*Old Statistical Account*, vol. xiii. p. 202). In

by coarse bread and cold water; and to Wicbert,[1] who, "multos annos in Hibernia peregrinus anchoreticam in magna perfectione vitam egerat."[2] Reginald of Durham has left a work on the life, penances, medical and other miracles, of the celebrated St. Godric, who, during the twelfth century, lived for about forty years as an anchorite in the hermitage of Finchale, on the river Weir, near Durham.[3] The same author speaks of, as contemporary holy hermits, St. Elric of Wlsingham, and an anchorite at Yareshale, on the Derwent.[4]* A succession of hermits occupied a cell near Norham.[5] Small islands appear to have been specially selected by the early anchorets for their heremitical retreats. Hereberct, the friend of St. Cuthbert, lived, according to Bede, an anchoret life upon one

Marmion (Canto i. 29) Sir Walter Scott describes the "Palmer" as, with solemn vows to pay,

"To fair St. Andrews bound,
Within the *ocean-cave* to pray,
Where good St. Rule his holy lay,
From midnight to the dawn of day,
Sung to the billows' sound."

[1] *Historia Ecclesiastica Gentis Anglorum*, lib. v. cap. 12.

[2] *Ibid.* lib. v. c. 9. Bede further states that this anchoret subsequently went to Frisland to preach as a missionary there, but he reaped no fruit from his labours among his barbarous auditors. "Returning then (adds Bede) to the beloved place of his peregrination, he gave himself up to our Lord in his wonted repose; for since he could not be profitable to strangers by teaching them the faith, he took care to be the more useful to his own people by the example of his virtue."

[3] Published in 1845 by the Surtees Society, *Libellus de Vita, etc., S. Godrici*, p. 65, etc.

[4] *Ibid.* pp. 45 and 192.

[5] *Ibid.* foot-note, p. 46.

* [See Wordsworth's beautiful inscription—"For the spot where the hermitage stood on St. Herbert's island, Derwentwater."—Ed. of 1858, p. 258. —P.]

of the islands in the lake of Derwentwater; and St. Cuthbert himself, Ethelwald, and Felgeld, when they aspired to the rank of anchoretish perfection (gradum anchoreticæ sublimitatis), successively betook themselves for this purpose to Farne, on the coast of Northumberland, a small isle about eight or nine miles south of Lindisfarne.[1] Among other anchorets who subsequently lived on Farne, Reginald incidentally mentions Aelric, Bartholomew, and Aelwin.[2] On Coquet Island, lying also off the Northumbrian coast, St. Henry the Dane led the life of a religious hermit, and died about the year 1120.[3] Inchcolm is not the only island in the Firth of Forth which is hallowed by the reputation of having been the residence of anchorets, seeking for scenes in which they might practise uninterrupted devotion. Thus, St. Baldred or Balther lived for some time, during the course of the seventh century, as a religious recluse, upon the rugged and precipitous island of the Bass, as stated by Boece, Leslie, Dempster,[4] etc., and, as we know with more certainty from a poem written—upwards now of one thousand years ago—by a native of this country, the celebrated Alcuin.[5] The followers of the order of St. Columba who desired

[1] Bede's *Vita Sancti Cuthberti*, cap. 16, 28, 46, etc.

[2] *De Beati Cuthberti Virtutibus*, pp. 63 and 66.

[3] See, *The Flowers of the Lives of the most renowned Saincts of the Three Kingdoms*, by Hierome Porter, p. 321.

[4] Boece's *History and Chronicles of Scotland*, book ix. c. 17, or vol. ii. p. 98; Leslie's *De Rebus Gestis Scotorum*, lib. iv. p. 152; Dempster's *Historia Ecclesiastica Gentis Scotorum*, lib. ii. p. 122, or vol. i. p. 66.

[5] The poem alluded to is designated "De Pontificibus et Sanctis Ecclesiæ Eboracencis." A copy of it is printed in Gale's *Historiæ Britannicæ, etc. Scriptores*, vol. iii. p. 703, *seq.* The famous author of this poem, Alcuin, who was brought up at York, and probably born there about the year 735, became afterwards, as is well known, the councillor and confidant of Charle-

to follow a more ascetic life than that which the society of his religious houses and monasteries afforded to its ordinary members,

magne. The application to the Bass of the lines in which he describes the anchoret residence of St. Balther is evident :

Est locus undoso circumdatus undique ponto,
Rupibus horrendis prærupto et margine septus,
In quo belli potens terreno in corpore miles
Sæpius aërias vincebat Balthere turmas ; etc.

The Bass was not the only hermit's island on our eastern coasts which was imagined, in these credulous times, to be the occasional abode of evil spirits. According to Bede no one had dared to dwell alone on the island of Farne before St. Cuthbert selected it as his anchoret habitation, because demons resided there (propter demorantium ibi phantasias demonum). *Vita Cuthberti*, cap. 16. See also the undevilling of the cave of Dysart by St. Serf in the foot-note of page 125, *supra*; and some alleged feats of St. Patrick and St. Columba in this direction in Dr. O'Donovan's *Annals of the Four Masters*, vol. i. p. 156. Two other islands in the Firth of Forth are noted in ancient ecclesiastical history—viz., Inch May and Inch Keith. "The ile of May, decorit (to use the words of Bellenden) with the blude and martirdome of Sanct Adriane and his fallowis," was the residence of that Hungarian missionary and his disciples when they were attacked and murdered about the year 874 by the Danes (Bellenden's *Translation of Boece's History*, vol. i. p. 37 ; see also vol. ii. p. 206 ; Dempster's *Historia Eccl. Gentis Scotorum*, lib. i. 17, and vol. i. p. 20 ; and Fordun, in the *Scotichronicon*, lib. i. c. vi., where he describes "Maya, prioratus cujus est cella canonicorum Sancti Andreæ de Raymonth ; ubi requiescit Sanctus Adrianus, cum centum sociis suis sanctis martyribus." Inch Keith is enumerated by Dr. Reeves (*Preface to Life of Columba*, p. 66) as one of the Scotch churches of St. Adamnan, Abbot of Iona from A.D. 679 to 704, and the biographer of St. Columba *—Fordun

* [As I have not the *Life of Columba* at hand to refer to, I must assume that so able an archæologist as my friend Dr. Reeves had sufficient authority for this statement. If it rested only on Fordun or Wynton, I should deem their authority insufficient to establish as a fact what seems to me so improbable. Assuming the story to have had a foundation, might not the real Adamnan have been the priest and monk of the monastery of Coludi or Coldingham, of whom Bede has written ? Coldingham, in his time, belonged to the Northumbrian kingdom.—P.]

sometimes withdrew (observes Dr. Reeves[1]) to a solitary place in the neighbourhood of the monastery, where they enjoyed undisturbed meditation, without breaking the fraternal bond. Such, in 634, was Beccan, the "solitarius," as he is designated in Cummian's

having long ago described it as a place "in qua præfuit Sanctus Adamnanus abbas, qui honorifice suscepit Sanctum Servanum, cum sociis suis, in ipsa insula, ad primum suum adventum in Scotiam." Andrew Wynton, himself the Prior of St. Serf's Isle in Lochlevin, describes also, in his old metrical *Orygynale Chronykil of Scotland*, vol. i. p. 128, this apocryphal meeting of the two saints

"at Inchkeith,
The ile betweene Kingorne and Leth."

The Breviary of Aberdeen, in alluding to this meeting, points out that the St. Serf received by Adamnan was not the St. Serf of the Dysart Cave, and hence also not the baptiser of St. Kentigern at Culross, as told in the legend of his mother, St. Thenew, or St. Thenuh—a female saint whose very existence the good Presbyterians of Glasgow had so entirely lost sight of, that centuries ago they unsexed the very name of the church dedicated to her in that city, and came to speak of it under the uncanonical appellation of St. Enoch's. This first St. Serf and Adamnan lived two centuries, at least, apart. In these early days Inch Keith was a place of no small importance, if it be—as some (see Macpherson's *Geographical Illustrations of Scottish History*) have supposed—the "urbs Giudi" of Bede, which he speaks of as standing in the midst of the eastern firth, and contrasts with Alcluith or Dumbarton, standing on the side of the western firth. The Scots and Picts were, he says, divided from the Britons "by two inlets of the sea (duobus sinibus maris) lying betwixt them, both of which run far and broad into the land of Britain, one from the Eastern, and the other from the Western Ocean, though they do not reach so as to touch one another. The eastern has in *the midst of it* the city of Giudi (Orientalis habit in medio sui *urbem Giudi*). The western has on it, that is, on the right hand thereof (ad dextram sui), the city of Alcluith, which in their language means the 'Rock of Cluith,' for it is close by the river of that name (Clyde)." (Bede's *Hist. Ecclesiast.*, book i. c. xii.) In reference to the supposed identification of Inch Keith and this "urbs Giudi," let me add (1.) that Bede's descrip-

[1] See his edition of Adamnan's *Life of Saint Columba*, p. 366.

contemporary Paschal letter to Segene, the Abbot of Iona; and such was Finan, the hermit of Durrow, who, in the words of Adamnan, "vitam multis anchoreticam annis irreprehensibiliter ducebat." According to the evidence of the Four Masters, an anchorite held the Abbacy of Iona in 747; another anchorite was Abbot-elect in 935; and a third was made Bishop in 964.* "The abode of such anchorites was (adds Dr. Reeves) called in Irish a 'desert' (Dysart), from the Latin *desertum;* and as the heremitical life was held in such honour among the Scotic Churches, we frequently find this word 'desert' an element in religious nomenclature. There was a 'desert' beside the monastery of Derry; and that belonging to Iona

tion (in medio sui) as strongly applies to the Island of Garvie, or Inch Garvie, lying midway between the two Queensferries: (2.) it is perhaps worthy of note that the term "Giudi" is in all probability a Pictish proper name, one of the kings of the Picts being surnamed "Guidi," or rather "Guidid" (see Pinkerton's *Inquiry into the History of Scotland*, vol. i. p. 287, and an extract from the *Book of Ballymote*, p. 504); and (3.) that the word "urbs," in the language of Bede, signifies a place important, not so much for its size as from its military or ecclesiastic rank, for thus he describes the rock (petra) of Dumbarton as the "urbs Alcluith," and Coldingham as the "urbs Coludi" (*Hist. Eccl.*, lib. iv. c. 19. etc.),—the Saxon noun "*ham*," house or village, having, in this last instance, been in former times considered a sufficient appellative for a place to which Bede applies the Latin designation of "urbs."

* [Colgan refers to the Life of *S. Fintani Eremitæ ad* 15 *Novemb., Tr. T.*, p. 606:—"Tir mille anachoritas in Momonia cum S. Hibaro Episcopo cujusdam quæstionis decidendæ causâ simul collectos, quos & Angelus Dei ad convivium à S. Brigida Christo paratum invitavit; invoco in auxilium per Jesum Christum." Quoted from the *Book of Litanies of S. Ængus*, on the same page.

See also the *Summary of the Saints* in that *Litany* in Ward's *Vita S. Rumoldi*, pp. 204, 205.

In short, the notices of deserts, hermits, and anchorites to be found, lives of saints, etc. etc., are innumerable.—P.]

was situate near the shore, in the low ground north of the Cathedral, as may be inferred from Port-an-discart, the name of a little bay in this situation." The charters of the Columbian House at Kells show that a "desert" existed in connection with that institution. Could the old building or capellula on Inchcolm have served as a "desert" to the Monastery there? *

The preceding remarks have spun out to a most unexpected extent; and I have to apologise both for their extravagant length and rambling character. At the same time, however, I believe that it would be considered an object of no small interest if it could be shown to be at all probable that we had still near us a specimen, however rude and ruinous, of early Scoto-Irish architecture. All authorities now acknowledge the great influence which, from the sixth to the eleventh or twelfth century, the Irish Church and Irish clergy exercised over the conversion and civilisation of Scotland. But on the eastern side of the kingdom we have no known remains of Scoto-Irish ecclesiastical architecture except the beautiful and perfect Round Tower of Brechin,[1] and the ruder

[1] This is no fit place to discuss the ages of the two Round Towers of Brechin and Abernethy. But it may perhaps prove interesting to some future antiquary if it is here mentioned, that when Dr. Petrie, in his *Ecclesiastical Architecture of Ireland* (p. 410), gives "about the year 1020" † as the probable

* [I think it very improbable, if the monastery founded by Alexander be meant.—P.]

† [The recollection of the error which I made by a carelessness not in such matters usual with me, in assigning this date 1020 instead of between the years 971 and 994, as I ought to have done, has long given me annoyance, and a lesson never to trust to memory in dates; for it was thus I fell into the mistake. I had the year 1020 on my mind, which is the year assigned by Pinkerton for the writing of the *Chron. Pictorum*, and, without

and probably older Round Tower of Abernethy. If, to these two instances, we dare to conjoin a specimen of a house or oratory of the same Scoto-Irish style, and of the same ancient period, such as

date of the erection of the Round Tower of Brechin, he chiefly relied—as he has mentioned to me, when conversing upon the subject,—for this approach to the era of its building, upon that entry in the ancient *Chronicon de Regibus Scotorum*, etc., published by Innes, in which it is stated that King Kenneth MacMalcolm, who reigned from A.D. 971 to A.D. 994, "tribuit magnam civitatem Brechne domino." (See the Chronicon in Innes' *Critical Inquiry*, vol. ii. p. 788.) The peculiarities of architecture in the Round Tower of Brechin assimilate it much with the Irish Round Towers of Donoughmore and Monasterboice, both of which Dr. Petrie believes to have been built in or about the tenth century. If we could, in such a question, rely upon the authority of Hector Boece, the Round Tower of Brechin is at least a few years older than the probable date assigned to it by Dr. Petrie. For, in describing the inroads of the Danes into Forfarshire about A.D. 1012, he tells us that these invaders destroyed and burned down the town of Brechin, and all its great church, except "*turrim quandam rotundam* mira arte constructam." (*Scotorum Historiæ*, lib. xi. 251, of Paris Edit. of 1526.)* This reference to the Round Tower of Brechin has escaped detection, perhaps because it has been omitted by Bellenden and Holinshed in their translations. No historical notices, I believe, exist, tending to fix in any probable way the exact age of the Round Tower of Abernethy ; but one or two circumstances bearing upon the inquiry are worthy of note. We are informed, both by the *Chronicon Pictorum* and by Bede, that in the eighth or ninth year of his reign, or about

stopping to remember or to refer, I took it for granted that it was the year of Kenneth's death, or rather his gift.—P.]

* [I congratulate you warmly on the discovery of this interesting and most valuable notice. Surely Boece could have had no object to serve by forging such a statement, nor had he such antiquarian knowledge as would have enabled him to forge a statement so consistent with the conclusion fairly to be drawn from the entry in the chronicle, and the characteristics of the architecture of the tower itself. It appears to me that no rational scepticism can in future be indulged as to the conclusion that the erection of this beautiful tower *must* be referred to the last quarter of the tenth century.—P.]

the Oratory on Inchcolm seems to me probably to be, we would have in such a specimen an addition of some moment to this limited and meagre list. Besides, it would surely not be uninterest-

A.D. 563, Brude, King of the Picts, embraced Christianity under the personal teaching of St. Columba. At Brude's death, in 586, Garnard succeeded, and reigned till 597 ; and he was followed by Nectan II., who reigned till 617. Fordun (*Scotichronicon*, lib. iv. cap. 12) and Wynton (book v. ch. 12), both state that King Garnard founded the collegiate Church of Abernethy ; and Fordun further adds that he had found this information in a chronicle of the Church of Abernethy itself, which is now lost ; "in quadam Chronica ecclesiæ de Abirnethy reperimus." But the register of the Priory of St. Andrews mentions Garnard's successor on the Pictish throne, Nectan II., as the builder of Abernethy, "hic ædificavit Abernethyn" (Innes' *Critical Inquiry*, p. 800). The probability is, that Garnard, towards the end of his reign, founded and commenced the building of the church establishment of Abernethy, and that it was concluded and consecrated in the early part of the reign of Nectan. The church was dedicated to St. Brigid ; and the *Chronicon Pictorum* (Innes' *Inquiry*, p. 778), in ascribing its foundation to Nectan I. (about A.D. 455) instead of Nectan II., commits a palpable anachronism, and very evident error, as St. Brigid did not die till a quarter of the next century had elapsed. (*Annals of the Four Masters* under the year 525 ; Colgan's *Trias Thaumaturga*, p. 619.) Again, according to the more certain evidence of Bede, another Pictish king, still of the name of Nectan (Naitanus Rex Pictorum), despatched messengers, about the year 710, to Ceolfrid, Abbot of Bede's own Northumbrian monastery of Jarrow, requesting, among other matters, that architects should be sent to him to build in his country a church of stone, according to the manner of the Romans (et architectos sibi mitti petiit, qui juxta morem Romanorum ecclesiam in lapide in gente ipsius facerent. (*Hist. Eccles.*, lib. v. c. xxi.) Forty years previously, St. Benedict or Biscop, the first Abbot of Jarrow, had brought there from Gaul, masons (cæmentarios) to build for him "ecclesiam lapideam juxta Romanorum morem." (See Bede's *Vita Beatorum Abbatum.*) Now it is probable that the Round Tower of Abernethy was not built in connection with the church established there by the Pictish kings at the beginning of the seventh century, for no such structures seem to have been erected in connection with Pictish churches in any other part of the Pictish kingdom ; and if at Abernethy, the capital of the Picts, a Round

ing could we feel certain that we have still standing, within eight or ten miles of Edinburgh, a building whose roof had covered the

Tower had been built in the seventh century of stone and lime, the Abbot of Jarrow would scarcely have been asked in the eighth century, by a subsequent Pictish king, to send architects to show the mode of erecting a church of stone in his kingdom. Nor is it in the least degree more likely that these ecclesiastic builders, invited by King Nectan in the early years of the eighth century, erected themselves the Round Tower of Abernethy; for the building of such towers was, if not totally unknown, at least totally unpractised by the ecclesiastic architects of England and France within their own countries.* The Scotic or Scoto-Irish race became united with the Picts into one kingdom in the year 843, under King Kenneth MacAlpine, a lineal descendant and representative of the royal chiefs who led the Dalriadic colony from Antrim to Argyleshire, about A.D. 506. (See the elaborate genealogical table of the Scottish Dalriadic kings in Dr. Reeves' edition of *Adamnan's Life of Columba*, p. 438.) The purely "Scotic period" of our history, as it has been termed, dates from this union of the Picts and Scots under Kenneth MacAlpine in 843, till Malcolm Canmore ascended the throne in 1057; and there is every probability that the Round Towers of Abernethy and Brechin were built during the period between these two dates, or during the regime of the intervening Scotic or Scoto-Irish kings,—in imitation of the numerous similar structures belonging to their original mother-church in Ireland. We

* [The determining the age of the Brechin tower—a question which I consider as now settled—must go far towards enabling us to come to a right conclusion as to the age of the tower of Abernethy; for I think that no one possessed of ordinary powers of observation and comparison, who has examined both, can for a moment doubt that the age of the Abernethy tower is much greater than that of the tower of Brechin. This is the opinion which I formed many years ago, after a very careful examination of the architectural peculiarities of each; and I came to the conclusion that the safest opinion which could be indulged as to the age of the Abernethy tower was, that it had been erected during the reign of the third Nectan, *i.e.* between 712 and 727, and by those Northumbrian architects of the monastery of Jarrow, for whose assistance that king, according to the high authority of Bede, had applied to build for him in his capital a stone church in the

head of King Alexander I., though it covered it for three days only; for that very circumstance would at the same time go far to esta-

may feel very certain, also, that they were not erected later than the commencement of the twelfth century, for by that date the Norman or Romanesque style,—which presents no such structures as the Irish Round Towers, was apparently in general use in ecclesiastic architecture in Scotland, under the pious patronage of Queen Margaret Atheling and her three crowned sons. Abernethy—now a small village—was for centuries a royal and pontifical city, and the capital of a kingdom, "fuit locus ille sedes principalis, regalis, et pontificalis, totius regni Pictorum" (Goodall's *Scotichronicon*, vol. i. p. 189); but all its old regal and ecclesiastical buildings have utterly vanished, with the exception only of its solitary and venerable Round Tower. And perhaps the preservation of the Round Tower in this, and in numerous instances in Ireland, amidst the general ruin and devastation which usually surround them, is owing to the simple circumstance that these Towers—whatever were their uses and objects—were structures which, in consequence of their remarkable combination of extreme tallness and slenderness, required to be constructed from the first of the very best and strongest, and consequently of the most durable building materials which could be procured; while the one-storeyed or two-storeyed wood-roofed churches, and other low and lighter ecclesiastical edifices with which they were associated, demanded far less strength in the original construction of their walls, and consequently have, under the dilapidating effects of centuries, much more speedily crumbled down and perished.

Roman style. In the features of that style, during the eighth century, as exhibited in its doorway, and, still more, its upper apertures, this tower appeared to stand alone—there is nothing similar to it to be seen either in Scotland or Ireland. The tower of Brechin has indeed a Romanesque doorway, but it is plainly of a later age, and its other features are quite Irish. The circumstance of the Abernethy doorway being placed on a level with the ground, and not, as almost universally, at a considerable height from it, seemed also to support this opinion, as it indicated that the erection of the tower was of a period anterior to the irruption of the Northmen, which rendered such a defensive feature an imperative necessity. I cannot agree with you in opinion as to the cause assigned for the preservation of the towers; for, in the first place, it is not true that their materials were stronger or

blish another fact, namely, that any such building might claim to be now the oldest roofed stone habitation in Scotland.*

better, or their construction in any way different from that of the churches with which they were connected, as proved by numerous examples in Ireland. Their walls are rarely found of greater thickness than those of their contemporaneous churches, where such have remained; and in all such cases the character of the masonry is identical. The cause which I should rather assign for this greater longevity would be their rotundity, and still more, their superior altitude. A church of moderate size, and humble height, might be easily injured, or even destroyed, by neighbouring or foreign assailants, but the destruction of a tower, or even its injury, beyond the burning of its wooden floors and doorway, would be a tedious and difficult labour, requiring ladders, with which we are not to suppose the incendiaries came provided; and hence their worst antagonist was found to be the flame from heaven. —P.]

* [Might not *oratory* be a safer term than *habitation?* Surely the clochans or monks' houses, called *stone pyramids* by Martin, in St. Kilda, and of which many are still perfect, are as old as Christianity in the *north* of Scotland, or as any similar buildings to be found in Ireland.—P.]

Fig. 13. Oratory on Inchcolm, as lately repaired by the Earl of Moray.

ON THE CAT-STANE, KIRKLISTON.

THE Mediæval Archæology of Scotland is confessedly sadly deficient in *written* documents. From the decline of Roman records and rule, onward through the next six or eight centuries, we have very few, or almost no written data to guide us in Scottish historical or antiquarian inquiries. Nor have we any numismatic evidence whatever to appeal to. In consequence of this literary dearth, the roughest lapidary inscriptions, belonging to these dark periods of our history, come to be invested with an interest much beyond their mere intrinsic value. The very want of other contemporaneous lettered documents and data imparts importance to the rudest legends cut on our ancient lettered stones. For even brief and meagre tombstone inscriptions rise into matters of historical significance, when all the other literary chronicles and annals of the men and of the times to which these inscriptions belong have, in the lapse of ages, been destroyed and lost.

It is needless to dwell here on the well-known fact, that in England and Scotland there have been left by the Roman soldiers and colonists who occupied our island during the first four centuries of the Christian era, great numbers of inscribed stones. British antiquarian and topographical works abound with descriptions and drawings of these Roman lapidary writings. But of late years another class or series of lapidary records has been particularly attracting the attention of British antiquaries,—viz., inscribed stones of a late Roman or post-Roman period. The inscriptions on this latter class of stones are almost always, if not always, sepulchral. The characteristically rude letters in which they are written consist —in the earliest stones—of debased Roman capitals; and—in the latest—of the uncial or minuscule forms of letters which are used in the oldest English and Irish manuscripts. Some stones show an intermixture of both alphabetical characters. These "Romano-British" inscribed stones, as they have been usually termed, have hitherto been found principally in Wales, in Cornwall, and in West Devon. In the different parts of the Welsh Principality, nearly one hundred, I believe, have already been discovered. In Scotland, which is so extremely rich in ancient sculptured stones, very few inscribed stones are as yet known; but if a due and diligent search be instituted, others, no doubt, will betimes be brought to light.

An inscribed Scottish stone of the class I allude to is situated in the county of Edinburgh, and has been long known under the name of the Cat-stane or Battle-stone. Of its analogy with the earliest class of Romano-British inscribed stones found in Wales, I was not fully aware till I had an opportunity of examining last year, at the meeting of the Cambrian Archæological Society, a valuable collection of rubbings and drawings of these Welsh stones, brought forward by that excellent antiquary, Mr. Longueville Jones; and afterwards, *in situ*, one or two of the stones themselves. I venture, in the following remarks, to direct the attention of the Society to the Cat-stane, partly in consequence of this belief in its analogy with the earliest Welsh inscribed stones; partly, also, in order to adduce an old and almost unknown description of the Cat-stane, made in the last years of the seventeenth century, by a gentleman who was perhaps the greatest antiquary of his day ; and partly because I have a new conjecture to offer as to the historical personage commemorated in the inscription, and, consequently, as to the probable age of the inscription itself.

Site and Description of the Stone.

The Cat-stane stands in the parish of Kirkliston, on the farm of Briggs,[1] in a field on the north side of the road to Linlithgow, and between the sixth and seventh milestone from Edinburgh. It is placed within a hundred yards of the south bank of the Almond ; nearly half-a-mile below the Boathouse Bridge ; and about three miles above the entrance of the stream into the Firth of Forth, at the old Roman station of Cramond, or Caer Amond. The monument is located in nearly the middle of the base of a triangular fork of ground formed by the meeting of the Gogar Water with the river Almond. The Gogar flows into the Almond about six or seven hundred yards below the site of the Cat-stane.[2] The ground on

[1] The farm is called " Briggs, or Colstane " (Catstane), in a plan belonging to Mr. Hutchison, of his estate of Caerlowrie, drawn up in 1797. In this plan the bridge (brigg) over the Almond, at the boathouse, is laid down. But in another older plan which Mr. H. has of the property, dated 1748, there is no bridge, and in its stead there is a representation of the ferry-boat crossing the river.

[2] In this strategetic angular fork or tongue of ground, formed by the confluence of these two rivers, Queen Mary and her suite were, according to Mr. Robert Chambers, caught when she was carried off by Bothwell on the 24th of April 1567. (See his interesting remarks " On the Locality of the Abduction of Queen Mary " in the *Proceedings of the Society of Scottish Antiquaries*, vol. ii. p. 331.)

which the Cat-stane stands is the beginning of a ridge slightly elevated above the general level of the neighbouring fields. The stone itself consists of a massive unhewn block of the secondary greenstone-trap of the district, many large boulders of which lie in the bed of the neighbouring river. In form it is somewhat prismatic, or irregularly triangular, with its angles very rounded. This

Fig. 14.

large monolith is nearly twelve feet in circumference, about four feet five inches in width, and three feet three inches in thickness. Its height above ground is about four feet and a half. The Honourable Mrs. Ramsay of Barnton, upon whose son's property the monument stands, very kindly granted liberty, last year, for an examination by digging beneath and around the stone. The accompanying woodcut is a copy of a sketch, made at the time, by my friend Mr. Drummond, of the stone as exposed when pursuing this search around its exposed basis. We found the stone to be a block seven feet three inches in total length, and nearly three feet buried in the soil. It was placed upon a basis of stones, forming apparently the remains of a built stone grave, which contained no bones[1] or other relics, and that had very evidently been already searched and harried. I shall indeed have immediately occasion to cite a passage proving that a century and a half ago the present pillar-stone was surrounded, like some other ancient graves, by a circular range of large flat-laid stones; and when this outer circle was removed,—if not before,—the vicinity and base of the central pillar were very probably dug into and disturbed.

Different Readings of the Inscription.

The inscription upon the stone is cut on the upper half of the

[1] The comparative rapidity or slowness with which bones are decomposed and disappear in different soils, is sometimes a question of importance to the antiquary. We all know that they preserve for many long centuries in dry soils and dry positions. In moist ground, such as that on which the Cat-stane stands, they melt away far more speedily. On another part of Mrs. Ramsay's property, namely in the policy, and within two hundred yards of the mansion-house of Barnton, I opened, several years ago, with Mr. Morritt of Rokeby, the grave of a woman who had died—as the tombstone on the spot told us—during the last Scottish plague in the year 1648. The only remains of sepulture which we found were some fragments of the wooden coffin, and the enamel crowns of a few teeth. All other parts of the body and skeleton had entirely disappeared. The chemical qualities of the ground, and consequently of its water, will of course modify the rapidity of such results.

eastern and narrowest face of the triangular monolith. Various descriptions of the legend have been given by different authors. The latest published account of it is that given by Professor Daniel Wilson in his work on *Scottish Archæology*. He disposes of the stone and its inscription in the two following short sentences :—"A few miles to the westward of this is the oft-noted Catt Stane in Kirkliston parish, on which the painful antiquary may yet decipher the imperfect and rudely-lettered inscription—the work, most probably, of much younger hands than those that reared the mass of dark whinstone on which it is cut—IN [H]OC TVMVLO IACET VETTA . . VICTR . . About sixty yards to the west of the Cat-stane a large tumulus formerly stood, which was opened in 1824, and found to contain several complete skeletons ; but nearly all traces of it have now disappeared."[1]

In the tenth volume of the *Statistical Account of Scotland*, collected by Sir John Sinclair, and published in 1794, the Rev. Mr. John Muckarsie, in giving an account of the parish of Kirkliston, alludes in a note to the "Cat-stane standing on the farm of that name in this parish." In describing it he observes "The form is an irregular prism, with the following inscription on the south-east face, deeply cut in the stone, in a most uncouth manner :—

IN OC T
VMVLO IACI
VETTA D
VICTA

We are informed," continues Mr. Muckarsie, "by Buchanan and other historians, that there was a bloody battle fought near this

[1] *Prehistoric Annals of Scotland*, p. 96.

place, on the banks of the Almond, in the year 995, between Kennethus, natural brother and commander of the forces of Malcolm II., King of Scotland, and Constantine, the usurper of that crown, wherein both the generals were killed. About two miles higher up the river, on the Bathgate road, is a circular mound of earth (of great antiquity, surrounded with large unpolished stones, at a considerable distance from each other, evidently intended in memory of some remarkable event). The whole intermediate space, from the human bones dug up, and graves of unpolished stones discovered below the surface, seems to have been the scene of many battles."[1]

In the discourse which the Earl of Buchan gave in 1780 to a meeting called together for the establishment of the present Society of Scottish Antiquaries, his Lordship took occasion to allude to the Cat-stane when wishing to point out how monuments, rude as they are, "lead us to correct the uncertain accounts which have been handed down by the monkish writers." "Accounts, for example, have (he observes) been given of various conflicts which took place towards the close of the tenth century between Constantine IV. and Malcolm, the general of the lawful heir of the Scottish Crown, on the banks of the River Almond, and decided towards its confluence to the sea, near Kirkliston. Accordingly, from Mid-Calder, anciently called Calder-comitis, to Kirkliston, the banks of the river are filled with the skeletons of human bodies, and the remains of warlike weapons; and opposite to Carlowrie there is a well-known stone near the margin of the river, called by the people *Catt Stane.* The

[1] *Statistical Account of Scotland*, collected by Sir John Sinclair, vol. x pp. 68, 75.

following inscription was legible on the stone in the beginning of this (the eighteenth) century; and the note of the inscription I received from the Rev. Mr. Charles Wilkie, minister of the parish of Ecclesmachan, whose father, Mr. John Wilkie, minister of the parish of Uphall, whilst in his younger days an inhabitant of Kirkliston, had carefully transcribed :—

IN HOC TUM · JAC · CONSTAN · VIC · VICT·"[1]

Lord Buchan adduces this alleged copy of the Cat-stane inscription as valuable from having been taken early in the last century. The copy of the inscription, though averred to be old, is, as we shall see in the sequel, doubtlessly most inaccurate. And there exist accounts of the inscription both older and infinitely more correct and trustworthy.

The oldest and most important notice of the Cat-stane and its inscription that I know of is published in a work where few would expect to find it—viz., in the *Mona Antiqua Restaurata* of the Rev. Mr. Rowlands. It is contained in a letter addressed to that gentleman by the distinguished Welsh archæologist, Edward Lhwyd. The date of Mr. Lhwyd's letter is "Sligo, March 12th, 1699-1700." A short time previously he had visited Scotland, and "collected a considerable number of inscriptions." At that time the Cat-stane was a larger and much more imposing monument than it is now, as shown in the following description of it. "One monument," says he, "I met with within four miles of Edinburgh, different from all I had seen elsewhere, and never observed by their antiquaries. I

[1] The *Scots Magazine* for 1780, p. 697. See also Smellie's *Account of the Institution and Progress of the Society of Antiquaries of Scotland* (1782), p. 8.

take it to be the tomb of some Pictish king; though situate by a river side, remote enough from any church. It is an area of about seven yards diameter, raised a little above the rest of the ground, and encompassed with large stones; all which stones are laid lengthwise, excepting one larger than ordinary, which is pitched on end, and contains this inscription in the barbarous characters of the fourth and fifth centuries, IN OC TUMULO JACIT VETTA F. VICTI. This the common people call the *Cat-Stene*, whence I suspect the person's

Fig. 15.

name was *Getus*, of which name I find three Pictish kings; for the names pronounced by the Britons with *G*, were written in Latin with *V*, as we find by Gwyrtheyrn, Gwyrthefyr, and Gwythelyn, which were written in Latin Vortigernus, Vortimerus, and Vitelinus."[1]

Besides writing the preceding note to Dr. Rowland regarding the Cat-stane, Mr. Lhwyd, at the time of his visit, took a sketch of the inscription itself. In the *Philosophical Transactions* for February 1700, this sketch of the Cat-stane inscription was, with eight others, published by Dr. Musgrave, in a brief communication entitled, "An Account of some Roman, French, and Irish Inscriptions and Antiquities, lately found in Scotland and Ireland, by Mr. Edward Lhwyd, and communicated to the publisher from Mr. John Hicks of Trewithier, in Cornwall." The accompanying woodcut (Fig. 15) is an exact copy of Mr. Lhwyd's sketch, as published in the *Philosophical Transactions.* In the very brief communication accompanying it, the Cat-stane is shortly described as "A Pictish monument near Edinburgh, IN OC TUMULO JACIT VETA F. VICTI. This the common people call the Ket-stean; note that the British names beginning with the letter Gw began in Latin with V [and the three examples given by Lhwyd in his letter to Dr. Rowland follow]. So I suppose (it is added) this person's name was Gweth or Geth, of which name were divers kings of the Picts, whence the vulgar name of Ketstone."[2]

In the course of the last century, notices or readings of the Cat-stane inscription, more or less similar to the account of it in the *Philosophical Transactions*, were published by different writers, as by

[1] Rowlands' *Mona Antiqua Restaurata*, second edition, p. 313. The inscription is printed in italics by Rowland. I have printed this and some of the following readings in small Roman capitals, in order to assimilate them all the more with each other.

[2] *Philosophical Transactions*, vol. xxii. p. 790.

Sir Robert Sibbald, in 1708,[1]—by Maitland, in 1753,[2]—by Pennant, in his journey through Scotland in 1772,[3]—and by Gough, in 1789, in the third volume of his edition of Camden's *Britannia*.[4]

All the four authors whom I have quoted agree as to the reading of the inscription, and give the two names mentioned in it, as VETTA and VICTI. But in printing the first of these names, VETTA, Maitland and Pennant, following perhaps the text in the *Philosophical Transactions*, carelessly spell it with a single instead of a double T; and Gough makes the first vowel in VICTI an E instead of an I. Sir Robert Sibbald gives as a K the mutilated terminal letter in the third line, which Mr. Llwyd deciphered as an F. Sibbald's account of the stone and its inscription, in 1708, is short but valuable, as affording an old independent reading of the legend. It is contained in his folio essay or work entitled, *Historical Inquiries Concerning the Roman Monuments and Antiquities in Scotland* (p. 50). "Close (says he) by Kirkliston water, upon the south side, there is a square pillar over against the Mannor of Carlowry with this inscription:—

[1] *Historical Inquiries concerning the Roman Monuments and Antiquities in Scotland*, p. 50.

[2] *The History of Edinburgh*, p. 508.

[3] *Tour in Scotland* in 1772, Part ii. p. 237. When describing his ride from Kirkliston to Edinburgh, he observes: "On the right hand, at a small distance from our road are some rude stones. On one called the *Cat-stean*, a compound of Celtic and Saxon, signifying the Stone of Battle, is this inscription: IN HOC TUMULO JACET VETA F. VICTI; supposed in memory of a person slain there."

[4] Camden's *Britannia*, edited by Richard Gough, vol. iii. p. 317. Mr. Gough cites also as Mr. Wilkie's reading, "IN HOC TUM, JAC. CONSTANTIE VICT."

IN OC TV
MVLO IACIT
VETTA K
VICTI

This (Sibbald continues) seemeth to have been done in later times than the former inscriptions [viz., those left in Scotland by the Romans]. Whether it be a Pictish monument or not is uncertain; the vulgar call it the CAT *Stane*."

Mr. Gough, when speaking of the stone in the latter part of the last century, states that the inscription upon it was "not now legible." It is certainly still even sufficiently legible and entire to prove unmistakably the accuracy of the reading of it given upwards of a century and a half ago by Lhwyd and Sibbald. The letters come out with special distinctness when examined with the morning sun shining on them; and indeed few ancient inscriptions in this country, not protected by being buried, are better preserved,—a circumstance owing principally to the very hard and durable nature of the stone itself, and the depth to which the letters have been originally cut. The accompanying woodcut is taken from a photograph of the stone by my friend Dr. Paterson, and very faithfully represents the inscription. The surface of the stone upon which the letters are carved has weathered and broken off in some parts; particularly towards the right-hand edge of the inscription. This process of disintegration has more or less affected the terminal letters of the four lines of the inscriptions. Yet, out of the twenty-six letters composing the legend, twenty are still comparatively entire and perfectly legible; four are more or less defective; and two nearly obliterated. The two which are almost obliterated consist of the first V in TVMVLO, constituting the terminal letter of the first

Fig. 16. The Cat-Stane, Kirkliston, *from a Photograph.*

line, and the last vowel I, or rather, judging from the space it occupies, E in JACIT. A mere impress of the site of the bars of the V is faintly traceable by the eye and finger, though the letter came out in the photograph. Only about an inch of the middle portion of the upright bar of I or E in JACIT can be traced by sight or touch. In this same word, also, the lower part of the C and the cross stroke of the T is defective. But even if the inscription had not been read when these letters were more entire, such defects in particular letters are not assuredly of a kind to make any palæographer entertain a doubt as to the two words in which these defects occur being TVMVLO and JACIT.

The terminal letter in the third line[1] was already defective in the time of Edward Lhwyd, as shown by the figure of it in his sketch. (See woodcut, No. 15.) Sibbald prints it as a K, a letter without any attachable meaning. Lhwyd read it as an F (followed apparently by a linear point or stop), and held it to signify—what F so often does signify in the common established formula of these old inscriptions—F(ILIVS). The upright limb of this F appears still well cut and distinct; but the stone is much hollowed out and destroyed immediately to the right, where the two cross bars of the letter should be. The site of the upper cross-bar of the letter is too much decayed and excavated to allow of any distinct recognition of it. The site, however, of a small portion of the middle cross bar is traceable at the point where it is still united to and springs from the upright limb of the letter. Beyond, or to the right of this letter F, a line about half-an-inch long, forming possibly a terminal stop or point of a linear type, commences on the level of the lower line of the letters, and runs obliquely upwards and outwards, till it is

[1] In the VETTA of this line the cross bar in A is wanting, from the stone between the upright bars being chipped or weathered out.

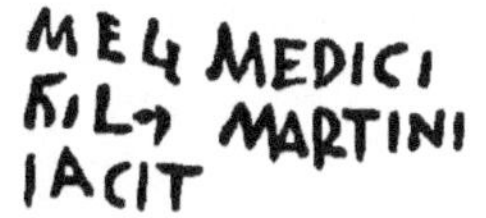

now lost above in the weathered and hollowed-out portion of stone. Its site is nearer the upright limb or basis of the F than it is represented to be in the sketch of Mr. Lhwyd, where it is figured as constituting a partly continuous extension downwards of the middle bar of the letter itself. And perhaps it is not a linear point, but more truly, as Lhwyd figures it, the lower portion of a form of the middle bar of F, of an unusual though not unknown type. The immediate descent or genealogy of those whom these Romano-British inscriptions commemorate is often given on the stones, but their status or profession is seldom mentioned. We have exceptions in the case of one or two royal personages, as in the famous inscription in Anglesey to "CATAMANUS, REX SAPIENTISSIMUS OPINATISSIMUS OMNIUM REGUM." The rank and office of priests are in several instances also commemorated with their names, as in the Kirkmadrine Stone in Galloway. In the churchyard of Llangian, in Caernarvonshire, there is a stone with an ancient inscription written not horizontally, but vertically (as is the case with regard to most of the Cornish inscribed stones), and where MELUS, the son of MARTINUS, the person commemorated, is a physician—MEDICVS. But the inscription is much more interesting in regard to our present inquiry in another point. For—as the accompanying woodcut of the Llangian inscription shows—the F in the word FILI is very much of the same type or form as the F seen by Lhwyd in the Cat-stane, and drawn by him. (See his sketch in the preceding woodcut, Fig. 15.) The context and position of this letter F in the Llangian legend leaves no doubt of its true character. The form is old; Mr. Westwood considers the age of the Llangian inscription as "not later than the fifth century."[1] An approach to the same form of F in the same word FILI, is seen in an inscribed stone which formerly stood at Pant y Polion in Wales, and is now removed to Dolan Cothy House. Again, in some instances, as in the Romano-British stones at Llandysilir, Clyddan, Lllandyssul, etc., where the F in Filius is tied to the succeeding I, the conjoined letters present an appearance similar to the F on the Cat-stane as figured by Lhwyd.

While all competent authorities are nearly agreed as to the lettering and reading of the first three lines, latterly the terminal letter of the fourth or last line has given rise to some difference of opinion. Lhwyd, Sibbald, and Pennant, unhesitatingly read the whole last line as VICTI. Lhwyd, in his sketch of the inscription, further shows that, following the last I, there is a stop or point of a linear form. The terminal I is three inches long, while the linear point or stop following it is fully an inch in length. Between it and the terminal I is a smooth space on the stone of five or six lines. Latterly this terminal I, with its superadded linear point, has been supposed by Mr. Muckarsie to be an A, and by Dr. Wilson to be an R. Both suppositions appear to me to be erroneous; and of this one or two considerations will, I think, satisfy any cautious observer who will examine carefully either the stone itself, or the cast of the inscription that was made in 1824—copies of which are placed in our own and in other museums. Mr. Muckarsie and Dr. Wilson hold the upright bar forming the letter I to be the primary upright bar of an A or R; and they think the remaining portions of these letters to be indicated or formed by the linear stop figured by Lhwyd.

[1] *Archæologia Cambrensis* (for 1848), vol. iii. p. 107.

That the letter is not A, is shown by the bar being quite perpendicular, and not oblique or slanting, as in the two other A's in the inscription. Besides, the middle cross stroke of the A is wanting; and the second descending bar of the letter is quite deficient in length—a deficiency not explicable by mutilation from the weathering of the stone, as the stone happens to be still perfectly entire both at the uppermost and the lowest end of this bar or line. This last reason is also in itself a strong if not a sufficient ground for rejecting the idea that the letter is an R; inasmuch as if it had been an R, the tail of the letter would have been found prolonged downwards to the base line of the other letters in the word. For it is to be held in remembrance, that though the forms of the letters in this inscription are rude and debased, yet they are all cut with firmness and fulness.

The idea that the terminal letter of the inscription is an R seems still more objectionable in another point of view. To make it an R at all, we can only suppose the disputed "line" to be the lowest portion of the segment of the loop or semicircular head of the R. The line, which is about an inch long, is straight, however, and not a part of a round curve or a circle, such as we know the mason who carved this inscription could and did cut, as witnessed by his O's and C's. Besides, if this straight line had formed the lower segment of the semicircular loop or head of an R, then the highest point of that R would have stood so disproportionately elevated above the top line or level of the other letters in this word, as altogether to oppose and differ from what we see in the other parts of this inscription. This same reason bears equally against another view which perhaps might be taken; namely, that the straight line in question is the tail or terminal right-hand stroke of the R, placed nearly horizontally, as is occasionally the form of this letter in some early inscriptions, like those of Yarrow and Llangian. But if this view be adopted, then the loop or semicircular head of the R must be considered as still more disproportionately displaced upwards above the common level of the top line; for in this view the whole loop or head must have stood entirely above this straight horizontal line, which line itself reaches above the middle height of the upright bar forming the I. Immediately above the horizontal line, for a space about an inch or more in depth, and some ten or twelve inches in length, there has been a weathering and chipping off of a splinter of the surface of the stone, as indicated by its commencement in an abrupt, curved, rugged edge above. This lesion or fracture of the stone has, I believe, originally given rise to the idea of the semblance of this terminal letter of the inscription to an R. Probably, also, this disintegration is comparatively recent; for in the last century Lhwyd, Sibbald, Maitland, and Pennant, all unhesitatingly lay down the terminal letter as an I. But even if it were an A or an R, and not an I and hyphen point, this would not affect or alter the view which I will take in the sequel, that the last word in the inscription is a Latinised form of the surname VICTA or WECTA; as, amid the numberless modifications to which the orthography of ancient names is subjected by our early chroniclers, the historic name in question is spelled by Ethelwerd with a terminal R,—in one place as UUITHAR, and in another as WITHER.[1] Altogether, however,

[1] See his "Chronicon," in the *Monumenta Historica Britannica*, pp. 502 and 505. Nouns, and names ending thus in "r," preceded by a vowel, were

I feel assured that the more accurately we examine the inscription as still left, and the more we take into consideration the well-known caution and accuracy of Edward Lhwyd as an archæologist, the more do we feel assured that his reading of the Cat-stane legend, when he visited and copied it upwards of a hundred and sixty years ago is strictly correct, viz.—

IN OC TV
MVLO JACIT
VETTA F.
VICTI.

Palæographic Peculiarities.

The palæographic characters of the inscription scarcely require any comment. As in most other Roman and Romano-British inscriptions, the words run into each other without any intervening space to mark their separation. The letters all consist of debased Roman capitals. They generally vary from two and a half to three inches in

often written without the penultimate vowel, particularly in the Scandinavian branches of the Teutonic language ; as Baldr for Balder and Baldur ; Folkvangr for Folkvangar ; Surtr for Surtur and Surtar, etc. (See the Glossary to the prose Edda in Bohn's edition of Mallet's *Northern Antiquities*, and Kemble's *Saxons in England*, pp. 346, 363, etc.) For genealogical lists full of proper names ending in "r" with the elision of the preceding vowel, see the long tables of Scandinavian and Orcadian pedigrees printed at the end of the work on the pre-Columbian discovery of America, *Antiquitates Americanæ*, etc., which was published at Copenhagen in 1837 by the Royal Society of Northern Antiquaries. In the first table of genealogies giving the pedigree of Thorfinn, the son of Sigurd, of the Orkney dynasty, etc., we have, among other names—Olafr, Grismr, Ingjaldr, Oleifr (*Rex Dublini*) ; Thorsteinn Raudr (*partis Scotiæ Rex*) ; Dungadr (*Earl of Katanesi*) ; Arfidr, Havadr, Thorfinnr, etc. (*Earls of Orkney*) ; etc. etc.

length; but the O in the first line is only one and a half inch deep. The O in TVMVLO in these ancient inscriptions is often, as in the Cat-stane, smaller than the other letters. M. Edmond Le Blant gives numerous marked instances of this peculiarity of the small O in the same words, "IN HoC TVMVLo," in his work on the early Romano-Gaulish inscriptions of France.[1] Most of the letters in the Cat-stane inscription are pretty well formed, and firmly though rudely cut. The oblique direction of the bottom stroke of L in TVMVLO is a form of that letter often observable in other old Romano-British inscriptions, as on the stone at Llanfaglan in Wales. The M in the same word has its first and last strokes splaying outwardly; a peculiarity seen in many old Roman and Romano-British monuments—as is also the tying together of this letter with the following V. In the Romano-British inscription upon the stone found at Yarrow, and which was brought under the notice of the Society by Dr. John Alexander Smith, there are three interments, as it were, recorded, the last of them in these words;[2]

...HIC IACENT
IN TVMVLO DVO FILI
LIBERALI.

The letters on this Yarrow stone are—with one doubtful except-

[1] *Inscriptions Chrétiennes de la Gaule, anterieures au VIII. Siècle.* See Plates Nos. 10, 11, 15, 16, 24, 25, etc.

[2] The name LIBERALIS is probably the Latinised form of a British surname having the same meaning. Rydderch, King of Strathclyde, in the latter part of the sixth century, and the personal friend of Kentigern and Columba, was sometimes, from his munificence, termed Rydderch *Hael*, or, in its Latinised form, Rydderch *Liberalis.* The first lines of the Yarrow inscription appear to me to read as far as they are decipherable, as follows:—

tion[1]—Roman capitals, of a ruder, and hence perhaps later, type than those cut on the Cat-stane; but the letters MV in TVMVLO are tied together in exactly the same way on the two stones. The omission of the aspirate in (H)OC, as seen on the Cat-stane, is by no means rare. The so-called bilingual, or Latin and Ogham, inscribed stone at Llanfechan, Wales, has upon it the Latin legend TRENACATVS IC JACET FILIVS MAGLAGNI—the aspirate being wanting in the word HIC. It is wanting also in the same way, and in the same word, in the inscription on the Maen Madoc stone, near Ystradfellte—viz., DERVACI FILIVS IVLII IC IACIT; and on the Turpillian stone near Crickhowel. In a stone, described by Mr. Westwood, and placed on the road from Brecon to Merthyr, the initial aspirate in "hoc" is not entirely dropped, but is cut in an uncial form, while all the other letters are Roman capitals; thus IN hOC TVMULO.

Linear hyphen-like stops, such as Lhwyd represents at the end of the fourth, and probably also of the third line on the Cat-stane inscription, seem not to be very rare. In the remarkable inscription on the Caerwys stone, now placed at Downing Whitford, "Here lies a good and noble woman"—[2]

HIC MEMOR IACIT F
LOIN : : : NI : : : : HIC
PE : : M
DVMNOGENI.

The true character of the G in the fourth line was first pointed out by Dr. Smith. It is of the same form as the G in the famous SAGRAMANVS stone, etc.

[1] The exception is the letter D in DVO, which verges to the uncial form

[2] In the inscription all the words are, as usual, run together, with the

HIC JACIT / MVLI
ER BONA NOBILI(S)

an oblique linear point appears in the middle of the legend, after the word JACIT. The linear stop on the Cat-stane inscription, at the end of the fourth line, is, as already stated, fully an inch in length, but it is scarcely so deep as the cuts forming the letters; and the original surface of the stone at both ends of this terminal linear stop is very perfect and sound, showing that the line was not extended either upwards or downwards into any form of letter. Straight or hyphen lines, at the end both of words—especially of the proper names—and of the whole inscriptions, have been found on various Romano-British stones, as on those of Margan (the Naen Llythyrog), Stackpole, and Clydau, and have been supposed to be the letter I, placed horizontally, while all the other letters in these inscriptions are placed perpendicularly. Is it not more probable that they are merely points? Or do they not sometimes, like tied letters, represent both an I and a stop?

Who is Commemorated in the Cat-stane Inscription?

In the account which Mr. George Chalmers gives of the Antiquities of Linlithgowshire in his *Caledonia*, there is no notice of the inscription on the Cat-stane taken; but, with a degree of vagueness of which this author is seldom guilty, he remarks, that this monolith "is certainly a memorial of some conflict and of *some* person."[1]

exception of the Jacit and Mulier, which are separated from each other by the oblique linear point. See a plate of the inscription in the *Archæologia Cambrensis* for 1855, p. 153.

[1] *Caledonia*, vol. ii. p. 844.

Is it not possible, however, to obtain a more definite idea of the person who is named on the stone, and in commemoration of whom it was raised?

In the extracts that have been already given, it has been suggested, by different writers whom I have cited, that the Cat-stane commemorates a Scottish king, Constantine IV., or a Pictish king, Geth. Let us first examine into the probability of these two suggestions.

1. Constantine?—In the olden lists of our Scottish kings, four King Constantines occur. The Cat-stane has been imagined by Lord Buchan and Mr. Muckarsie to have been raised in memory of the last of these—viz., of Constantine IV., who fell in a battle believed by these writers to have been fought on this ground in the last years of the tenth century, or about A.D. 995. In the *New Statistical Account of Scotland*, the Reverend Mr. Tait, the present minister of Kirkliston, farther speaks of the "Catstean (as) supposed to be a corruption of Constantine, and to have been erected to the honour of Constantine, one of the commanders in the same engagement, who was there slain and interred."[1]

In the year 970 the Scottish king Culen died, having been "killed (according to the Ulster Annals), by the Britons in open battle;" and in A.D. 994, his successor, Kenneth MacMalcolm, the founder of Brechin, was slain.[2] Constantine, the son of Culen,

[1] New *Statistical Account of Scotland*, vol. i. p. 138. For the same supposed corruption of the name Constantine into Cat-stane, see also Fullarton's *Gazetteer of Scotland*, vol. ii. p. 182.

[2] The brief history of Kenneth, his parentage, reign, and mode of death, as given in one of the earliest Chronicles of the Kings of Scotland, quoted

reigned for the next year and a half, and fell in a battle for the crown fought between him and Kenneth, the son of Malcolm I. The site of this battle was, according to most of our ancient authorities, on the Almond. There are two rivers of this name in Scotland, one in Perthshire and the other in the Lothians. George Chalmers places the site of the battle in which Constantine fell on the Almond in Perthshire; Fordun, Boece, and Buchanan place it on the Almond in the Lothians, upon the banks of which the Cat-stane stands. The battle was fought, to borrow the words of the Scotichronicon, "in Laudonia juxta ripam amnis Almond."[1] *The Chronicle of Melrose* gives (p. 226) the "Avon"—the name of another large stream in the Lothians—as the river that was the site of the battle in question. Wynton (vol. i. p. 182) speaks of it as the "Awyne." Bishop Leslie transfers this same fight to the banks of the Annan in Dumfriesshire, describing it as having occurred during an invasion of Cumbria, "ad Annandiæ amnis ostia."[2]

Among the authorities who speak of this battle or of the fall of Constantine, some describe these events as having occurred at the source, others at the mouth of the Almond or Avon. Thus the ancient rhyming chronicle, cited in the Scotichronicon, gives the

by Father Innes (p. 802), contains in its few lines a very condensed and yet powerful story of deep maternal affection and fierce female revenge. The whole entry is as follows:—"Kinath Mac-Malcolm 24, an. et 2. mens. Interfectus in Fotherkern a suis per perfidium Finellæ filiæ Cunechat comitis de Angus; cujus Finellæ filium unicum prædictus Kinath interfecit apud Dunsinoen." The clumsy additions of some later historians only spoil and mar the original simplicity and force of this "three-volume" historical romance.

[1] Tom. i. p. 219, of Goodall's edition.

[2] *De Rebus Gestis Scotorum*, chap. lxxxi. p. 200.

locality of Constantine's fall as "ad caput amnis Amond."[1] *The Chronicle of Melrose*, when entering the fall of "Constantinus Calwus," quotes the same lines, with such modifications as follows :[2]—

> " Rex Constantinus, Culeno filius ortus,
> Ad caput amnis Avon ense peremtus erat,
> In Tegalere ; regens uno rex et semis annis,
> Ipsum Kinedus Malcolomida ferit."

Wyntown cites the two first of these Latin lines, changing, as I have said, the name of the river to Awyne, almost, apparently, for the purpose of getting a vernacular rhyme, and then himself tells us, that

> " At the Wattyr hed of Awyne,
> The King Gryme slwe this Constantyne." [3]

If the word "Tegalere" in the *Melrose Chronicle* be a true reading,[4] and the locality could be identified under the same or a similar derivative name, the site of the battle might be fixed, and the point ascertained whether it took place, as the preceding authorities aver, at the source, "water-head" or "caput" of the river ; or, as Hector, Boece and George Buchanan[5] describe it, at its mouth or entrance into the Forth at Cramond ; "ad Amundæ amnis ostia tribus passuum millibus ab Edinburgo."[6] A far older and far more valu-

[1] *Joannis Forduni Scotichronicon*, tom. i. p. 219.

[2] *Chronicon de Mailros*, p. 226 (Bannatyne Club edition).

[3] Wyntown's *Orygynale Cronykil of Scotland*, vol. i. p. 183.

[4] In the *Scotichronicon* instead of "In Tegalere," the third of these lines commences "Inregale regens," etc. ; and it is noted that in the "Liber Dumblain" the line begins "Indegale," etc.

[5] Buchanan, in his "*Rerum Scoticorum Historia*, gives the locality as "ad Almonis amnis ostium." (Lib. vi. c. 81.)

[6] *Scotorum Historiæ*, p. 235 of Paris edition of 1574. Bellenden and

able authority than either Bocce or Buchanan, namely, the collector of the list of the Scottish and Pictish kings, extracted by Sir Robert Sibbald from the now lost register of the Priory of St. Andrews,[1] seems also to place the death of King Constantine at the mouth of the Almond, if we interpret aright the entry in it of "interfectus in Rathveramoen" as meaning "Rath Inver Amoen,"—the rath or earth-fortress at the mouth of the Amoen.[2]

Even, however, were it allowed that the battle in which Constantine perished was fought upon the Almond, and not upon the Avon, on the stream of the former name in the Lothians and not in Perthshire, at the mouth and not at the source of the river, there still, after all, remains no evidence whatever that the Cat-stane

Stewart, in their translations of Bocce's *History* both place the fight at "Crawmond."

[1] This document, entitled *Nomina Regum Scottorum et Pictorum* and published by Father Innes in his *Critical Essay*, p. 797, etc., is described by that esteemed and cautious author as a document the very fact of the registration of which among the records and charters of the ancient church of St. Andrews "is a full proof of its being held authentick at the time it was written, that is about A.D. 1251." (P. 607.)

[2] The orthography of the copy of this Chronicle, as given by Innes, is very inaccurate, and the omission of the two initial letters of "*in*ver," not very extraordinary in the word Rathveramoen. Apparently the same word Rathinveramon occurs previously in the same Chronicle, when Donald MacAlpin, the second king of the combined Picts and Scots, is entered as having died "in Raith in Veramont" (p. 801). In another of the old Chronicles published by Innes, this king is said to have died in his palace at "Belachoir" (p. 783). If, as some historians believe, the Lothians were not annexed to Scotland before his death in A.D. 859, by Kenneth the brother of Donald, and did not become a part of the Scottish kingdom till the time of Indulf (about A.D. 954), or even later, then it is probable that the site of King Donald's death in A.D. 863, at Rathinveramon, was on the Almond in Perthshire, within his own territories.

was raised in commemoration of the fall of the Scottish king; whilst there is abundant evidence to the contrary. The very word "Inver," in the last of the designations which I have adduced, is strongly against this idea. For the term "Inver," when applied to a locality on a stream, almost invariably means the mouth of it,[1] and not a site on its course—such as the Cat-stane occupies—three miles above its confluence. Nor is there any probability that an inscribed monument would be raised in honour of a king who, like Constantine, fell in a civil war,—who was the last of his own branch of the royal house that reigned,—and was distinguished, as the ancient chroniclers tell us, by the contemptuous appellation of *Calvus.* There is great reason, indeed, to believe that the idea of the Cat-stane being connected with the fall of Constantine is comparatively modern in its origin. Oral tradition sometimes creates written history; but, on the other hand, written history sometimes creates oral tradition. And in the present instance a knowledge of the statements of our ancient historians in all probability gave rise to such attempts as that of Mr. Wilkie—to find, namely, a direct record of Constantine in the Cat-stane inscription. But when we

[1] I am only aware of one very marked exception to this general law Malcolm Canmore is known to have been killed near Alnwick, when attacking its castle. Alnwick is situated on the Alne, about five or six miles above the village of Alnmouth, the ancient Twyford, on the Alne, of Bede, on the mount near which St. Cuthbert was installed as a bishop. But in the ancient Chronicle from the Register of St. Andrews, King Malcolm is entered (see Innes, p. 803) as "interfectus in Inneraldan." The error has more likely originated in a want of proper local knowledge on the part of the chronicler than in so unusual a use of the Celtic word "inver;" for, according to all analogies, while the term is applicable to Alnmouth, it is not at all applicable to Alnwick.

compare the inscription itself, as read a century and a half ago by Lhwyd and Sibbald, and as capable of being still read at the present day, with the edition of it as given by Lord Buchan, it is impossible not to conclude that the idea of connecting the legend with the name of Constantine is totally without foundation. For, besides minor errors in punctuation and letterings, such as the total omission in Lord Buchan's copy of the inscription of the three last letters VLO of "TVMVLO," the changing of VETTA to VIC, etc., we have the two terminal letters of JACIT—viz. the IT, changed into the seven-lettered word CONSTAN, apparently with no object but the support of a theory as to the person commemorated in the legend and the monolith. Most assuredly there is not the very slightest trace of any letters on the surface of the stone where the chief part of the word CONSTAN is represented as existing—viz., after JACIT. It would be difficult, perhaps, to adduce a case of more flagrant incorrectness in copying an inscription than Mr. Wilkie's and Lord Buchan's reading of the Cat-stane legend affords. Mr. Gough, in his edition of Camden's *Britannia* (1784), only aggravates this misrepresentation. For whilst he incorrectly states that the inscription is "not now legible," he carelessly changes Mr. Wilkie's alleged copy of the leading word from CONSTAN to CONSTANTIE, and suppresses altogether the word VIC.

Getus, Gweth, or Geth?—I have already cited Mr. Lhwyd's conjecture that the Cat-stane is "the tomb of some Pictish King," and the opinion expressed by him and Mr. Hicks, that taking the V in the Latin VETTA of the inscription as equal to the Pictish letters G or Gw, the name of the Pictish king commemorated by the stone was Getus, "of which name," observes Mr. Lhwyd, "I find three Pictish kings." In the analogous account sent by Mr. Hicks to the *Philosophical Transactions* along with Mr. Lhwyd's sketch of the Cat-stane, it is stated that the person's name on "this Pictish monument" was Gweth or Geth, "of which name," it is added, "were divers kings of the Picts, whence the vulgar name of Ket-stone."

It is unnecessary to stop and comment on the unsoundness of this reasoning, and the improbability—both as to the initial and terminal letters—of the surname VETTA in this Latin inscription being similar to the Pictish surname Geth or GETUS, as Lhwyd himself gives and writes it in its Latin form. Among the lists of the Pictish kings, whilst we have several names beginning with G, we have some also commencing in the Latinised forms of the Chronicles with V, as Vist, Vere, Vipoignamet, etc.

But a much more important objection exists against the conjecture of Mr. Lhwyd, in the fact that his memory had altogether misled him as to there having been "three" Pictish kings of the name of "Getus," or "divers kings of the Picts of the names of Geth or Gweth," to use the words employed in the *Philosophical Transactions.*

Lists, more or less complete, of the Pictish kings have been found in the Histories of Fordun and Winton, in the pages of the Scalacronica and Chronicles of Tighernach, in the Irish copy of Nennius, in the extracts published by Sir Robert Sibbald and Father Innes from the lost Register of St. Andrews, and in the old Chronicum Regum Pictorum, supposed to be written about A.D. 1020, and preserved in the Colbertine Library.

None of these lists include a Pictish king of the name of Getus, Geat, or Gweth. Some of the authorities—as the Register of St. Andrews, Fordun, and Winton—enter as the second king of the Picts Ghede or Gede, the Gilgidi of the *Chronicum Regum Pictorum*; and this latter chronicle contains in its more mythical and earlier part the appellations Got, Gedeol, Guidid, and Brude-Guith; but none of these surnames sufficiently correspond either to Mr. Lhwyd's statement or to the requirements of the inscription.

But whilst thus setting aside the conjectures as to the Cat-stane commemorating the name of a Scottish King Constantine, or of a Pictish King Geth, I would further remark that the surname in the inscription, namely—VETTA FILIUS VICTI—is one which appears to me to be capable of another and a more probable solution. With this view let us proceed then to inquire who was

VETTA, *the son* of VICTUS?

And *first*, I would beg to remark, that the word Vetta is still too distinct upon the Cat-stane to allow of any doubt as to the mere name of the person commemorated in the inscription upon it.

Secondly, The name of Vetta, or, to spell the word in its more common Saxon forms, Wetta or Witta, is a Teutonic surname. To speak more definitely, it pertains to the class of surnames which characterised these so-called Saxon or Anglo-Saxon invaders of our island, and allied Germanic tribes, who overran Britain upon the decline of the Roman dominion amongst us.

Bede speaks, as is well known, of our original Teutonic conquerors in the fifth century as coming from three powerful tribes of Germany; namely, the Saxons, Angles, and Jutes. "Advenenerunt autem de tribus Germaniæ populis fortioribus, id est, Saxonibus, Anglis, Jutis" (lib. i. c. 20).[1] Ubo Emmius, in his *History of the Frisians*, maintains that "more colonies from Friesland than Saxony, settled in Briton, whether under the names of Jutes, or of Angles, or later of the Saxons."[2] Procopius, who lived nearly two centuries before Bede, and had access to good means of information from being the secretary of the Emperor Belisarius, states that at the time of his writing (about A.D. 548) three numerous nations possessed Britain, the Angles and Frisians (Αγγιλοι τε και φρισσονες), and those surnamed, from the Island, Brittones.[3] Modern Friesland seems to have yielded a considerable number of our Teutonic invaders and colonists; and it is in that isolated country that we find, at all events, the characteristics and language of our Teutonic forefathers best preserved. In his *History of England during the Anglo-Saxon Period*, the late Sir Francis Palgrave remarks, "The tribes by whom Britain was invaded, appear principally to have proceeded from the country now called Friesland. Of all the continental dialects (he adds), the ancient Frisick is the one which approaches most nearly to the Anglo-Saxon of our ancestors."[4] "The nearest approach," according to Dr. Latham, "to our genuine and typical German or Anglo-Saxon forefathers, is not to be found within the four seas of Britain, but in the present Frisian or Friesland."[5] At present, about one hundred thousand inhabitants of

[1] *Historia Ecclesiastica Gentis Anglorum*. (Stevenson's Edit. p. 35.)

[2] *De Bello Gothico*, lib. iv. c. 20. See other authorities in Turner's Anglo-Saxons, vol. i. p. 182.

[3] *Emmii Rerum Friescarum Historia*, p. 41.

[4] *History of England*, vol. i.—Anglo-Saxon Period, pp. 33, 34.

[5] The *Ethnology of the British Islands*, p. 259. At p. 240, Dr. Latham

Friesland speak the ancient or Country-Friesic, a language unintelligible to the surrounding Dutch, but which remains still nearly allied to the old Anglo-Saxon of England. Some even of their modern surnames are repetitions of the most ancient Anglo-Saxon surnames in our island, and, among others, still include that of Vetta or Witta; thus showing its Teutonic origin. In discussing the great analogies between ancient Anglo-Saxon and modern Friesic, Dr. Bosworth, the learned Professor of Anglo-Saxon Literature at Oxford, incidentally remarks, "I cannot omit to mention that the leaders of the Anglo-Saxons bear names which are now in use by the Friesians, though by time a little altered or abbreviated. They have Horste, Hengst, WITTE, Wiggele, etc., for the Anglo-Saxon Horsa, Hengist, WITTA, Wightgil, etc."[1]

states, "A native tradition makes Hengist a Frisian." Dr. Bosworth cites (see his *Origin of the English, etc., Language and Nation*, p. 52) Maerlant in his Chronicle as doubtful whether to call Hengist a Frisian or a Saxon.

[1] See his *Origin of the English, German, and Scandinavian Languages*, p. 54. Some modern authorities have thought it philosophical to object to the whole story of Hengist and Horsa, on the alleged ground that these names are "equine" in their original meaning—"henges" and "hors" signifying stallion and horse in the old Saxon tongue. If the principles of historic criticism had no stronger reasons for clearing the story of the first Saxon settlement in Kent of its romantic and apocryphal superfluities, this argument would serve us badly. For some future American historian might, on a similar hypercritical ground, argue against the probability of Columbus, a Genoese, having discovered America, and carried thither (to use the language of his son Ferdinand) "the olive branch and oil of baptism across the ocean,"—of Drake and Hawkins having, in Queen Elizabeth's time, explored the West Indies, and sailed round the southernmost point of America,—of General Wolfe having taken Quebec,—or Lord Lyons being English ambassador to the United States in the eventful year 1860, on the ground that Colombo is actually the name of a dove in Italian, Drake and Hawkins only the appellations of birds, and Wolfe and Lyons the English names for two wild beasts.

But Witta or Vetta was not a common name among our more leading Anglo-Saxon forefathers. Among the many historical surnames occurring in ancient Saxon annals and English chronicles, the name of Vetta, as far as I know, only occurs twice or thrice.

I. It is to be found in the ancient Saxon poem of *The Scop*, or *Traveller's Tale*, where, among a list of numerous kings and warriors, Vetta or Witta is mentioned as having ruled the Swaefs—

"Witta weold Swæfum."[1]

The Swaefs or Suevi were originally, as we know from classical writers, a German tribe, or confederacy of tribes, located eastward of the old Angles; and Ptolemy indeed includes these Angles as a branch of the Suevi. But possibly the Swaefs ruled by Wittan, and mentioned in *The Scop* in the preceding line, and in others (see lines 89 and 123), were a colony from this tribe settled in England.

II. In the list of the ancient Anglo-Saxon Bishops of Lichfield, given by Florence of Worcester, the name "Huita" occurs as tenth on the roll.[2] Under the year 737, Simeon of Durham enters the consecration of this bishop, spelling his name as Hweicca and Hweitta.[3] In a note appended to Florence's Chronicle, under the year 775, his death is recorded, and his name given as Witta.[4]

III. The name Vetta occupies a constant and conspicuous place in the lineage of Hengist and Horsa, as given by Bede, Nennius, the Saxon Chronicle, etc. In the list of their pedigree, Vetta or Witta is always represented as the grandfather of the Teutonic brothers.

[1] See Thorpe's edition of Beowulf and other Anglo-Saxon Poems, p. 219, line 45.

[2] *Monumenta Historica*, p. 623. [3] *Ib.*, p. 659. [4] *Ib.*, p. 544.

The inscription on the Cat-stane further affords, however, a most important *additional element* or criterion for ascertaining the particular Vetta in memory of whom it was raised; for it records the name of his father, namely, Victus or Victa. And in relation to the present inquiry, it is alike interesting and important to find that in the genealogy given by our ancient chronicles of the predecessors of Hengist and Horsa, whilst Vetta is recorded as their grandfather, Victi or Wecta is, with equal constancy, represented as their great-grandfather. The old lapidary writing on the Cat-stane describes the Vetta for whom that monument was raised as the son of Vecta; and the old parchment and paper writings of our earliest chroniclers invariably describe the same relationship between the Vetta and Victa of the forefathers of Hengist and Horsa. Thus Bede, when describing the invasion of England by the German tribes in the time of Vortigern, states that their "leaders were two brothers, Hengist and Horsa, who were the sons of Victgils, whose father was Vitta, whose father was Vecta, whose father was Woden, from whose stock the royal race of many provinces deduces its origin," "Erant autem filii Victgilsi, cujus pater Vitta, cujus pater Vecta cujus pater Voden, de cujus stirpe multarum provinciarum regum genus originem duxit."[1] In accordance with a common peculiarity in his orthography of proper names, and owing also, perhaps, to the character of the Northumbrian dialect of the Anglo-Saxon tongue, Bede spells the preceding and other similar surnames with an initial

[1] *Historia Ecclesiastica Gentis Anglorum*, lib. i. cap. 15, p. 34 of Mr. Stevenson's edition. In some editions of Bede's *History* (as in Dr. Giles' Translation, for example) the name of Vitta is carelessly omitted, as a word apparently of no moment. Such a discussion as the present shows how wrong it is to tamper with the texts of such old authors.

V, while by most other Anglo-Saxon chroniclers, and in most other Anglo-Saxon dialects, the surnames are made to commence with a W. Thus, the Vilfrid, Valchstod, Venta, etc., of Bede,[1] form the Wilfrid, Walchstod, Wenta (Winchester), etc., of other Saxon writers. In this respect Bede adheres so far to the classic Roman standard in the spelling of proper names. Thus, for example, the Isle of Wight, which was written as Wecta by the Saxons, is the Vecta and Vectis of Ptolemy and Eutropius, and the Vecta also of Bede; and the name Venta, just now referred to as spelled so by Bede, is also the old Roman form of spelling that word, as seen in the *Itinerary* of Antonine.

The *Saxon Chronicle* gives the details of the first advent of the Saxons under Hengist and Horsa in so nearly the same words as the *Historia Ecclesiastica*, as to leave no doubt that this, like many other passages in the earlier parts of the *Saxon Chronicle*, were mere translations of the statements of Bede. But most copies of the *Saxon Chronicle* were written in the dialect of the West Saxons, and, consequently, under A.D. 449, they commence the surnames in the pedigree of our Saxon invaders with a W,— as Wightgils, Witta, Wecta, etc.; telling us that Hengist and Horsa, "waeron Wihtgilses suna, Wihtgils waes Witting, Witta Wecting, Wecta Wodning," etc.

Ethelwerd, an Anglo-Saxon nobleman, who himself claimed to be a descendant of the royal stock of Woden, has left us a Latin history or Chronicle, "nearly the whole of which is an abridged translation of the *Saxon Chronicle*, with a few trivial alterations

[1] See these names in page 414 of Stevenson's edition of the *Historia Ecclesiastica*."

and additions."[1] In retranslating back into Latin, the Anglo-Saxon names in the genealogy of Hengist and Horsa, he makes the Wecta of the *Saxon Chronicle* end with an R,—a matter principally of interest because, as we have already seen, some have supposed the corresponding name in the Cat-stane to terminate with an R. Speaking of Hengist as leader of the Angles[2] Ethelwerd describes his pedigree thus:—"Cujus pater fuit Wihtgels avus Wicta; proavus WITHER, atavus Wothen," etc. In a previous page,[3] the same author tells us that Hengest et Horsa filii Uuyrhtelsi, avus eorum Uuicta, et proavus eorum Uuithar, atavus eorum Uuothen, qui est rex multitudinis barbarorum."

In the preceding paragraphs we find the same authors, or at least the scribes who copied their writings, spelling the same names in very diverse ways. All know how very various, and sometimes almost endless, is the orthography of proper nouns and names among our ancient chroniclers, and among our mediæval writers and clerks also. Thus Lord Lindsay, in his admirable *Lives of the Lindsays*, gives examples of above a hundred different ways in which he has found his own family name spelled. In the *Historia Britonum*, usually attributed to Nennius, the pedigree of the Saxon invaders of Kent is given at greater length than by Bede; for it is traced back four or five generations beyond Woden[4] up to Geat, and the spelling of the four races from Woden to Hengist and

[1] *Monumenta Historica Britt.*, preface, p. 82.

[2] "Ethelwerdi Chronicorum," lib. ii. c. 2, in *Monumenta Historica*, p. 505

[3] *Ibid.* lib. i. p. 502 of *Monumenta Historica.*

[4] The historical personage and leader Woden is represented in all these genealogies as having lived four generations, or from 100 to 150 years earlier than the age of Hengist and Horsa.

Horsa is varied according to the Celtic standard of orthography, as cited already from Edward Lhwyd—namely, the Latin and Saxon initials V and W are changed to the Cymric or British G, or GU. In the same way, the Isle of Wight, "Vecta" or "Wecta," is spelled in Nennius "Guith" and "Guied;" Venta (Winchester) is written Guincestra; Vortigernus, Guorthigernus; Wuffa, king of the east Angles, Guffa; etc. etc. In only one, as far as I am aware, of the old manuscript copies of the *Historia Britonum*, is the pedigree of Hengist and Horsa spelled as it is by Bede and all the Saxon writers, with an initial V or W, as Wictgils, Witta, Wecta, and Woden. This copy belongs to the Royal Library in Paris, and the orthography alone sufficiently determines it to have been made by an Anglo-Saxon scribe or editor. Of some twenty-five or thirty other known manuscripts of the same work, most, if not all, spell the ancestors of Hengist with the initial Keltic GU,—as "Guictgils, Guitta, Guechta"—one, among other arguments, for the belief that the original and most ancient part of this composite *Historia* was penned, if not, as asserted in many of the copies, by Gildas, a Strathclyde Briton, at least by a British or Cymric hand. The account given in the work of the arrival of the Saxons is as follows:—"Interea venerunt tres ciulæ a Germania expulsæ in exilio, in quibus erant Hors et Hengist, qui et ipsi fratres erant, filii Guictgils, filii Guitta, filii Guechta, filii Vuoden, filii Frealaf, filii Fredulf, filii Finn, filii Folcwald, filii Geta, qui fuit, aiunt filius Dei. Non ipse est Deus Deorum Amen, Deus exercitum, sed unus est ab idolis eorum quæ ipsi colebant."[1] In this pedigree of the ancestors of

[1] See p. 24 of Mr. Stevenson's edition of *Nennii Historia Britonum*, printed for the English Historical Society. In the Gaelic translation of the *Historia*

Hengist and Horsa, it is deserving of remark that Woden, from whom the various Anglo-Saxon kings of England, and other kings of the north-west of Europe generally claimed their royal descent, is entered as a historical personage, living (according to the usual reckoning applied to genealogies) about the beginning of the third century, and who could count his descent back to Geat; while the Irish and other authorities affect to trace his pedigree for some generations even beyond this last-named ancestor.[1] According to Mallet, the true name of this great conqueror and ruler of the north-western tribes of Europe was "Sigge, son of Fridulph; but he assumed the name of Odin, who was the supreme god among the Teutonic nations, either to pass, among his followers, for a man inspired by the gods, or because he was chief priest, and presided

Britonum, known as the Irish Nennius, the name Wetta or Guitta is spelled in various copies as "Guigte" and "Guite." The last form irresistibly suggests the Urbs Guidi of Bede, situated in the Firth of Forth. Might not he have thus written the Keltic or Pictish form of the name of a city or stronghold founded by Vitta or Vecta; and does this afford any clue to the fact, that the waters of the Forth are spoken of as the Sea of Guidi by Angus the Culdee, and as the Mare Fresicum by Nennius, while its shores are the Frisicum Litus of Joceline? In the text I have noted the transformation of the analogous Latin name of the Isle of Wight, "Vecta," into "Guith," by Nennius. The "urbs Guidi" of Bede is described by him as placed in the middle of the Firth of Forth, "in medio sui." Its most probable site is, as I have elsewhere (see *Proceedings of Society of Antiquaries of Scotland*, vol. ii. pp. 254, 255) endeavoured to show, Inch Keith; and, phonetically, the term "Keith" is certainly not a great variation from "Guith" or "Guidi." At page 7 of Stevenson's edition of Nennius, the Isle of Wight, the old "Insula Vecta" of the Roman authors, is written "Inis Gueith"—a term too evidently analogous to "Inch Keith" to require any comment.

[1] See Irish Nennius, p. 77; *Saxon Chronicle*, under year 855, etc.

over the worship paid to that deity."[1] In his conquering progress towards the north-west of Europe, he subdued, continues Mallet, "all the people he found in his passage, giving them to one or other of his sons for subjects. Many sovereign families (he adds) of the north are said to be descended from those princes." And Hengist and Horsa were thus, as was many centuries ago observed by William of Malmesbury, "the great-great-grandsons of that Woden from whom the royal families of almost all the barbarous nations derive their lineage, and to whom the Angles have consecrated the fourth day of the week (Wodens-day), and the sixth unto his wife Frea (Frey-day), by a sacrilege which lasts even *to this time*."[2]

Henry of Huntingdon, in his *Historiæ Anglorum*, gives the pedigree of Hengist and Horsa according to the list which he found in Nennius ; but he changes back the spelling to the Saxon form. They were, he says, "Filii Widgils, filii Wecta, filii Vecta, filii Woden, filii Frealof, filii Fredulf, filii Fin, filii Flocwald, filii Ieta (Geta)." Florence of Worcester follows the shorter genealogy of Bede, giving in his text the names of the ancestors of Hengist and Horsa as Wictgils, Witta, and Wecta ; and in his table of the pedigrees of the kings of Kent spelling these same names Wihtgils, Witta, and Wehta.[3]

In giving the ancient genealogy of Hengist and Horsa, we thus find our old chroniclers speaking of their grandfather under the

[1] *Northern Antiquities*, Bohn's edition, p. 71. Sigge is generally held as the name of one of the sons of Woden.

[2] *Gest.* I. sec. 5, I. 11.

[3] *Monumenta Historica Britannica*, p. 707.

various orthographic forms of Guitta, Uuicta, Witta, Vitta; and their great-grandfather as Guechta, Uuethar, Wither, Wechta, Wecta, and Vecta. In the Cat-stane inscription the last—Vecta or Victa—is placed in the genitive, and construed as a noun of the second declension, whilst Vetta retains, as a nominative, its original Saxon form. The older chroniclers frequently alter the Saxon surnames in this way. Thus, Horsa is sometimes made, like Victa, a noun of the second declension, in conjunction with the use of Hengist, Vortimer, etc., as unaltered nominatives. Thus, Nennius tells us,[1] "Guortemor cum Hengist et Horso . . . . pugnabat." (cap. xlvi.) According to Henry of Huntingdon, "Gortimer . . . . ex obliquo aciem Horsi desrupit," etc. (Lib. ii.)

The double and distinctive name of "Vetta filius Victa," occurring, as it thus does, in the lineage of Hengist and Horsa, as given both (1) in our oldest written chronicles and (2) in the old inscription carved upon the Cat-stane, is in itself a strong argument for the belief that the same personage is indicated in these two distinct varieties of ancient lettered documents. This inference, however, becomes still stronger when we consider the rarity of the appellation Vetta, and the great improbability of there having ever existed two historic individuals of this name both of them the sons of two Victas. But still, it must be confessed, various arguments naturally spring up in the mind against the idea that in the Cat-stane we have a memorial of the grandfather of Hengist and Horsa. Let us look at some of these reasons, and consider their force and bearing.

[1] See his "Chronicon ex Chronicis," in the *Monumenta Historica*, pp. 523 and 627.

Some Objections considered.

Perhaps, as one of the first objections, I should notice the doubts which some writers have expressed as to such leaders as Hengist and Horsa having ever existed, and as to the correctness, therefore, of that genealogy of the Saxon kings of Kent in which Hengist and Horsa are included.[1]

The two most ancient lists of that lineage exist, as is well known, in the "Historia Britonum" of Gildas or Nennius, and in the "Historia Ecclesiastica" of Bede.

The former of these genealogical lists differs from the latter in being much longer, and in carrying the pedigree several generations beyond the great Teutonic leader Woden, backwards to his eastern forefather, Geat, whom Mr. Kemble and others hold to have been probably the hero Woden, whose semi-divine memory the northern tribes worshipped. Both genealogical lists agree in all their main particulars back to Woden—and so far corroborate the accuracy of each other. Whence the original author of the *Historia Britonum* derived his list, is as unknown as the original authorship of the work itself. Some of Bede's sources of information are alluded to by himself. Albinus, Abbot of St. Augustine's, Canterbury, and Nothhelm, afterwards Archbishop of Canterbury, "appear," observes Mr. Stevenson, "to have furnished Bede with chronicles in which

[1] See preceding note (1), p. 168. In answer to the vague objection that the alleged leaders were two brothers, Mr. Thorpe observes that the circumstance of two brothers being joint-kings or leaders, bearing, like Hengist and Horsa, alliterative names, is far from unheard of in the annals of the north; and as instances (he adds) may be cited, Ragnar, Inver, Ulba, and two kings in Rumedal—viz. Haerlang and Hrollang.—See his Translation of Lappenberg's *History of the Anglo-Saxons*, vol. i. pp. 78 and 275.

he found accurate and full information upon the pedigrees, accessions, marriages, exploits, descendants, deaths and burials of the kings of Kent."[1] That the genealogical list itself is comparatively accurate, there are not wanting strong reasons for believing. The kings of the different seven or eight small Anglo-Saxon kingdoms of England all claimed—as the very condition and charter of their regality—a direct descent from Woden, through one or other of his several sons. To be a king among our Anglo-Saxon forefathers, it was necessary, and indeed indispensable, both to be a descendant of Woden, and to be able to prove this descent. The chronicles of most ancient people, as the Jews, Irish, Scots, etc., show us how carefully the pedigree of their royal and noble families was anciently kept and retained. And surely there is no great wonder in the Saxon kings of Kent keeping up faithfully a knowledge of their pedigree—say from Bede's time, backwards, through the nine or ten generations up to Hengist, or the additional four generations up to Woden. The wonder would perhaps have been much greater if they had omitted to keep up a knowledge, by tradition, poems, or chronicles, of a pedigree upon which they, and the other kings of the Saxon heptarchy, rested and founded—as descendants of Woden—their whole title to royalty, and their claim and charter to their respective thrones.[2]

[1] See Mr. Stevenson's Introduction, p. xxv., to the Historical Society's edition of Bede's *Historia Ecclesiastica*; and also Mr. Hardy in the Preface, p. 71, to the *Monumenta Historica Britannica*.

[2] The great importance attached to genealogical descent lasted much longer than the Saxon era itself. Thus the author of the latest Life (1860) of Edward I., when speaking of the birth of that monarch at London in 1239, observes (p. 8), "The kind of feeling which was excited by the birth of an

But a stronger objection against the idea of the Cat-stane being a monument to the grandfather of Hengist and Horsa rises up in the question,—Is there any proof or probability that an ancestor of Hengist and Horsa fought and fell in this northern part of the island, two generations before the arrival of these brothers in Kent?

It is now generally allowed, by our best historians, that before the arrival of Hengist and Horsa in Kent, Britain was well known at least to the Saxons and Frisians, and other allied Teutonic tribes.

Perhaps from a very early period the shores and comparative riches of our island were known to the Teutons or Germans inhabiting the opposite continental coast. "It seems hardly conceivable," observes Mr. Kemble, "that Frisians who occupied the coast (of modern Holland) as early as the time of Cæsar, should not have

English prince in the English metropolis, and by the king's evident desire to connect the young heir to the throne with his Saxon ancestors, is shown in the *Worcester Chronicle* of that date. The fact is thus significantly described:—

'On the 14th day of the calends of July, Eleanor, Queen of England gave birth to her eldest son Edward; whose father was Henry; whose father was John; whose father was Henry; whose mother was Matilda the Empress; whose mother was Matilda, Queen of England; whose mother was Margaret, Queen of Scotland; whose father was Edward; whose father was Edmund Ironside; who was the son of Ethelred; who was the son of Edgar; who was the son of Edmund; who was the son of Edward the elder; who was the son of Alfred.'"—(*The Greatest of the Plantagenets*, pp. 8 and 9.)

Here we have eleven genealogical ascents appealed to from Edward to Alfred. The thirteen or fourteen ascents again from Alfred to Cerdic, the first Anglo-Saxon king of Wessex, are as fixed and determined as the eleven from Alfred to Edward. (See them quoted by Florence, Asser, etc.) But the power of reckoning the lineage of Cerdic up through the intervening nine alleged ascents to Woden, was indispensable to form and to maintain Cerdic's claim to royalty, and was probably preserved with as great, if not greater care when written records were so defective and wanting.

found their way to Britain."[1] We know from an incident referred to by Tacitus, in his Life of Agricola, that at all events the passage in the opposite direction from Britain to the north-west shores of the Continent was accidentally revealed—if not, indeed, known long before—during the first years of the Roman conquest of Scotland. For Tacitus tells us that in A.D. 83 a cohort of Usipians, raised in Germany, and belonging to Agricola's army, having seized some Roman vessels, sailed across the German Ocean, and were seized as pirates, first by the Suevi and afterwards by the Frisians (*Vita Agricolæ*, xlv. 2, and xlvi. 2). In Agricola's Scottish army there were other Teutonic or German conscripts. According to Tacitus, at the battle of the Mons Grampius three cohorts of Batavians and two cohorts of Tungrians specially distinguished themselves in the defeat of the Caledonian army. Various inscriptions by these Tungrian cohorts have been dug up at Cramond, and at stations along the two Roman walls, as at Castlecary and Housesteads. At Manchester, a cohort of Frisians seems to have been located during nearly the whole era of the Roman dominion.[2] Another cohort of Frisian auxiliaries seems, according to Horsley, to have been stationed at Bowess in Richmondshire.[3] Teutonic officers were occasionally attached to other Roman corps than those of their own countrymen. A Frisian citizen, for example, was in the list of officers of the Thracian cavalry at Cirencester.[4] The

[1] *The Saxons in England*, vol. i. p. 11.

[2] See the inscription, etc., in Whittaker's *Manchester*, vol. i. p. 160.

[3] On these Frisian cohorts, and consequently also Frisian colonists, in England, see the learned *Memoir on the Roman Garrison at Manchester*, by my friend Dr. Black. (Manchester, 1849.)

[4] Buckman and Newmarch's work on *Ancient Corinium*, p. 114.

celebrated Carausius, himself a Menapian, and hence probably of Teutonic origin, was, before he assumed the emperorship of Britain, appointed by the Roman authorities admiral of the fleet which they had collected for the purpose of repressing the incursions of the Franks, Saxons, and other piratical tribes, who at that date (A.D. 287) ravaged the shores of Britain and Gaul.[1]

In the famous Roman document termed "Notitia utriusque Imperii," the fact that there were Saxon settlers in England before the arrival of Hengist and Horsa seems settled, by the appointment of a "Comes Littoris Saxonici in Britannica."[2] The date of this official and imperial Roman document is fixed by Gibbon between A.D. 395 and 407. About forty years earlier we have—what is more to our present purpose—a notice by Ammianus Marcellinus of Saxons being leagued with the Picts and Scots, and invading the territories south of the Forth, which were held by the Romans and their conquered allies and dependants—the Britons.

To understand properly the remarks of Ammianus, it is necessary to remember that the two great divisional military walls which the Romans erected in Britain, stretched, as is well known, entirely across the island—the most northerly from the Forth to the Clyde, and the second and stronger from the Tyne to the Solway. The large tract of country lying between these two military walls formed from time to time a region, the possession of which seems to have been debated between the Romans and the more northerly tribes; the Romans generally holding the country up to the northern wall or

[1] Palgrave's *Anglo-Saxons*, p. 24.

[2] For fuller evidence on this point, see the remarks by Mr. Kemble in his *Saxons in England*, vol. i. p. 13, etc.

beyond it, and occasionally being apparently content with the southern wall as the boundary of their empire.

About the year A.D. 369, the Roman general Theodosius, the father of the future emperor of the same name, having collected a disciplined army in the south, marched northward from London, and after a time conquered, or rather reconquered, the debateable region between the two walls ; erected it into a fifth British province, which he named "Valentia," in honour of Valens, the reigning emperor ; and garrisoned and fortified the borders (*limites* que vigiliis tuebatur et practenturis).[1] The notices which the excellent contemporary historian, Ammianus Marcellinus, has left us of the state of this part of Britain during the ten years of active rebellion and war preceding this erection of the province of Valentia are certainly very brief, but yet very interesting. Under the year 360, he states that "In Britain, the stipulated peace being broken, the incursions of the Scots and Picts, fierce nations, laid waste the grounds lying next to the boundaries (loca *limitibus* vicina vastarent)." "These grounds were," says Pinkerton, "surely those of the future province of Valentia."[2] Four years subsequently, or in 364, Ammianus again alludes to the Britons being vexed by continued attacks from the same tribes, namely the Picts and Scots, but he describes these last as now assisted by, or leagued with, the Attacots and with the *Saxons*

[1] *Ammiani Marcellini Historiæ*, lib. xxviii. c. 1. The poet Claudian, perhaps with the full liberty of a poet, sings of Theodosius' forces in this war having pursued the Saxons to the very Orkneys :—

——— maduerunt Saxone fuso
Orcades.

[2] *Inquiry into the History of Scotland*, vol. i. p. 116. See also Gibbon's *Decline and Fall*, chap. xxv.

—"Picti, SAXONESQUE, et Scotti, et Attacotti, Britannos aerumnis vexavere continuis." Again, under the year 368, he alludes to the Scots and Attacots still ravaging many parts; but now, instead of speaking of them as leagued with the Picts and Saxons, he describes them as combined with the Picts, divided into two nations, the Dicaledonæ and Vecturiones:—"Eo tempore Picti in duas gentes divisi, Diacaledonæ et Vecturiones, itidemque Attacotti, bellicosa hominum natio, et Scotti per diversa vagantes, multa populabuntur."

In both of these two last notices for the years 364 and 368, the invaders are described as consisting of four different tribes. The Scots and Attacots are mentioned under these appellations in both. But whilst, in the notice for 364, the two remaining assailants are spoken of as Picts and Saxons (Picti, Saxonesque), in the notice for 368 the remaining assailants are described as the "Picts, divided into the Dicaledonæ and Vecturiones." Is it possible that the Saxon allies were now amalgamated with the Picts, and that they assumed the name of Vecturiones after their leader Vetta or Vecta? The idea, at all events, of naming nations patronymically from their leaders or founders was common in ancient times, though the correctness of some of the instances adduced is more than doubtful. Early Greek and Roman history is full of such alleged examples; as the Trojans from Tros; the Achæans from Achæus; the Æolians from Æolus; the Peloponnesians from Pelops; the Dorians from Dorus; the Romans from Romulus, etc. etc.; and so is our own. The Scots from Ireland are, observes Bede, named to this day Dalreudins (Dalriads), from their commander Reuda.[1] The Irish

[1] *Histor. Eccles.*, lib. i. c. 1, § 8.

called (according to some ancient authorities) the Picts "Cruithne," after their alleged first king, Crudne or Cruthne. In a still more apocryphal spirit the word Britons was averred by some of the older chroniclers to be derived from a leader, Brito—"Britones Bruto dicti," to use the expression of Nennius (§ 18); Scots from Scota ("Scoti ex Scota," in the words of the *Chronicon Rythmicum*), etc.

The practice of eponymes was known also, and followed to some extent among the Teutonic tribes, both in regard to royal races and whole nations. The kings of Kent were known as Aescingas, from Aesc, the son of Hengist;[1] those of East Anglia were designated Wuffingas, after Wuffa ("Uffa, a quo reges Orientalium Anglorum Vuffingus appellant"[2]). In some one or other of his forms, Woden (observes Mr. Kemble) "is the eponymus of tribes and races. Thus, as Geat, or through Geat, he was the founder of the Geatas; through Gewis, of the Gewissas; through Scyld, of the Scyldingas, the Norse Skjoldungar; through Brand, of the Brodingas; perhaps, through Baetwa, of the Batavians."[3] It could therefore scarcely be regarded as very exceptional at least, if Vetta, one of the grandsons of Woden, should have given, in the same way, his name to a combined tribe of Saxons and Picts, over whom he had been elected as leader.[4]

[1] Bede's *Hist. Eccles.*, lib. ii. cap. v. (Oisc, a quo reges Cantuariorum solent Oiscingas cognominare.) [2] *Ibid.*, lib. ii. cap. xv.

[3] *The Saxons in England*, vol. i. p. 341.

[4] In his account of the kings of the Picts, Mr. Pinkerton (*Inquiry into History of Scotland*, vol. i. p. 293) calculates that the sovereign "Wradech Vechla" of the *Chronicon Pictorum* reigned about A.D. 380. In support of his own philological views, Mr. Pinkerton alters the name of this Pictish king from "Wradech Vechla" to "Wradech *Vechta*." There is not, however, I believe, any real foundation whatever for this last reading, interesting as it might be, in our present inquiry, if true.

That a Saxon force, like that mentioned by Ammianus as being joined to the Picts and Scots in A.D. 364, was led by an ancestor of Hengist and Horsa is quite in accordance with all that is known of Saxon laws and customs. As in some other nations, the leaders and kings were generally, if not always, selected from their royal stock. "Descent" (observes Mr. Kemble) "from Heracles was to the Spartans what descent from Woden was to the Saxons—*the* condition of royalty."[1] All the various Anglo-Saxon royal families that, during the time of the so-called Heptarchy, reigned in different parts of England certainly claimed this descent from Woden. Hengist and Horsa probably led the band of their countrymen who invaded Kent, as members of this royal lineage; and a royal pre-relative or ancestor would have a similar claim and chance of acting as chief of that Saxon force which joined the Picts and Scots in the preceding century.

If we thus allow, for the sake of argument, that Vetta, the son of Victus, the grandfather of Hengist and Horsa, is identical with Vetta the son of Victus commemorated in the Cat-stane inscription, and that he was the leader of those Saxons mentioned by Ammianus that were allied with the Picts in A.D. 364, we shall find nothing incompatible in that conjecture with the era of the descent upon Kent of Hengist and Horsa. Bede, confusing apparently the arrival of Hengist and Horsa with the date of the second instead of the first visit of St. Germanus to Britain, has placed at too late a date the era of their first appearance in Kent, when he fixes it in the year 449. The facts mentioned in the earlier editions or copies of Nennius have led our very learned and accurate colleague Mr.

[1] *The Saxons in England*, vol. i. p. 149.

Skene, and others, to transfer forwards twenty or more years the date at which Hengist and Horsa landed on our shores.[1] But whether Hengist and Horsa arrived in A.D. 449, or, as seems more probable, about A.D. 428, if we suppose them in either case to have been born about A.D. 400, we shall find no incongruity, but the reverse, in the idea that their grandfather Vetta was the leader of a Saxon force thirty-six years previously. Hengist was in all probability past the middle period of life when he came to the Court of Vortigern, as he is generally represented as having then a daughter, Rowena, already of a marriageable age.

On the cause or date of Vetta's death we have of course no historical information; but the position of his monument renders it next to a certainty that he fell in battle; for, as we have already seen, the Cat-stane stands, in the words of Lhwyd, "situate on a river side, remote enough from any church." The barrows and pillar stones placed for miles along that river prove how frequently it had served as a strategic point and boundary in ancient warfare.[2] The field in which the Cat-stane itself stands was, as we have already found Dr. Wilson stating, the site formerly of a large tumulus. In a field, on the opposite bank of the Almond, my friend, Mr. Hutchison of Caerlowrie, came lately, when prosecuting some draining

[1] Mr. Hardy, in the preface (p. 114, etc.) to the *Monumenta Historica Britannica*, maintains also, at much length, that the advent and reception of the Saxons by Vortigern was in A.D. 428, and not 449. He contests for an earlier Saxon invasion of Britain in A.D. 374. See also Lappenberg in his *History of England under the Anglo-Saxon Kings*, vol. i. pp. 62, 63.

[2] Two miles higher up the river than the Cat-stane, four large monoliths still stand near Newbridge. They are much taller than the Cat-stane, but contain no marks or letters on their surfaces. Three of them are placed around a large barrow.

operations on his estate, upon numerous stone-kists, which had mutual gables of stone, and were therefore, in all probability, the graves of those who had perished in battle. Whether the death of Vetta occurred during the war with Theodosius in A.D. 364, or, as possibly the appellation Vecturiones tends to indicate, at a later date, we have no ground to determine.

The vulgar name of the monument, the Cat-stane, seems, as I have already hinted, to be a name synonymous with Battle-stane, and hence, also, so far implies the fall of Vetta in open fight. Maitland is the first author, as far as I am aware, who suggests this view of the origin of the word Cat-stane. According to him, "Catstean is a Gaelic and English compound, the former part thereof (Cat) signifying a battle, and stean or stan a stone ; so it is the battlestane, in commemoration probably of a battle being fought at or near this place, wherein Veta or Victi, interred here, was slain."[1] I have already quoted Mr. Pennant, as taking the same view of the origin and character of the name; and Mr. George Chalmers, in his *Caledonia*, propounds the same explanation of the word :—"In the parish of Liberton, Edinburghshire, there were (he observes) several large cairns, wherein were found various stone chests, including urns, which contained ashes and weapons; some of these cairns which still remain are called the *Cat*-stanes or Battle-stanes.[2] Single stones in various parts of North Britain are still known under the appropriate name of *Cat*-stanes. The name (he adds) is plainly derived from the British *Cad*, or the Scoto-Irish *Cath*, which

[1] *History of Edinburgh*, p. 509.

[2] *Transactions of the Society of Scottish Antiquaries*, vol. i. p. 308. Maitland, in his *History of Edinburgh*, p. 307, calls these cairns the "Cat-heaps."

signify a battle."[1] But the word under the form *Cat* is Welsh or British, as well as Gaelic. Thus, in the *Annales Cambriæ*, under the year 722, the battle of Pencon is entered as "Cat-Pencon."[2] In his edition of the old Welsh poem of the Gododin, Williams (verse 38) prints the battle of Vannau (Manau) as "Cat-Vannau."

The combination of the Celtic word "Cat" with the Saxon word "stane" may appear at first as an objection against the preceding

[1] *Caledonia*, vol. i. p. 86. The only references, however, which Mr. Chalmers gives to a "single stone" in Scotland, bearing the name of Cat-stane, all relate to this monument in Kirkliston parish :—"The tallest and most striking ancient monolith in the vicinity of Edinburgh is a massive unhewn flat obelisk, standing about ten feet high, in the parish of Colinton." Maitland (*History of Edinburgh*, p. 507), and Mr. Whyte (*Trans. of Scottish Antiquaries*, vol. i. p. 308) designate this monument the Caiy-stone. "Whether this (says Maitland) be a corruption of the Catstean I know not." The tall monolith is in the neighbourhood of the cairns called the Cat-stanes or Cat-heaps (see preceding note). Professor Walker, in an elaborate Statistical Account of the Parish of Colinton, published in 1808, in his *Essays on Natural History* describes the Cat-heaps or cairns as having been each found, when removed, to cover a coffin made of *hewn* stones. In the coffins were found mouldering human bones and fragments of old arms, including two bronze spear-heads. "When the turnpike road which passes near the above cairns was formed, for more than a mile the remains of dead bodies were everywhere thrown up." Most of them had been interred in stone coffins made of coarse slabs. To use the words of Professor Walker, "Not far from the three cairns is the so called 'Caiy-stone' of Maitland and Whyte. It has always, however (he maintains), been known among the people of the country by the name of the Ket-stane." It is of whinstone, and "appears not to have had the chisel, or any inscription upon it." "The craig (he adds) or steep rocky mountain which forms the northern extremity of the Pentland Hills, and makes a conspicuous figure at Edinburgh, hangs over this field of battle. It is called Caer-*Ketan* Craig. This name appears to be derived from the Ket-stane above described, and the fortified camp adjacent, which, in the old British, was termed a Caer." (P. 611.)

[2] See "Annales Cambriæ," in the *Monumenta Hist. Britannica*, p. 833.

idea of the origin and signification of the term Cat-stane. But many of our local names show a similar compound origin in Celtic and Saxon. In the immediate neighbourhood, for example, of the Cat-stane,[1] we have instances of a similar Celtic and Saxon amalgamation in the words Gogar-burn, Lenny-bridge, Craigie-hill, etc. One of the oldest known specimens of this kind of verbal alloy, is alluded to above a thousand years ago by Bede,[2] in reference to a locality not above fourteen or fifteen miles west from the Cat-stane. For, in his famous sentence regarding the termination of the walls of Antoninus on the Forth, he states that the Picts called this eastern "head of the wall" Pean-fahel, but the Angles called it Pennel-*tun*. To a contracted variety of this Pictish word signifying head of the wall, or to its Welsh form Pengual, they added the Saxon word "town," probably to designate the "villa," which, according to an early addition to Nennius, was placed there. "Pengaaul, quæ villa Scottice Cenail [Kinneil], Anglice verò Peneltun dicitur."[3]

[1] In Maitland's time (1753), there was a farm-house termed "Catstean," standing near the monument we are describing. And up to the beginning of the present century the property or farm on the opposite side of the Almond, above Caerlowrie, was designated by a name, having apparently the Celtic "battle" noun as a prefix in its composition—viz., Cat-elbock. This fine old Celtic name has latterly been changed for the degenerate and unmeaning term Almond-hill.

[2] *Historia Ecclesiast.*, lib. i. c. xii. "Sermone Pictorum Peanfahel, lingua autem Anglorum Penneltun appellatur."

[3] *Historia Britonum*, c. xix. At one time I fancied it possible that the mutilated and enigmatical remains of ancient Welsh poetry furnished us with a name for the Cat-stane older still than that appellation itself. Among the fragments of old Welsh historical poems ascribed to Taliesin, one of the best known is that on the battle of Gwen-Ystrad. In this composition the poet describes, from professedly personal observation, the feats at the above battle

The palæographic peculiarities of the inscription sufficiently bear out the idea of the monument being of the date or era which I have ventured to assign to it—a point the weight and importance of which it is unnecessary to insist upon. "The inscription," says Lhwyd, "is in the barbarous characters of the fourth and fifth centuries." Professor Westwood, who is perhaps our highest

of the army of his friend and great patron, Urien, King of Rheged, who was subsequently killed at the siege of Medcaut, or Lindisfarne, about A.D. 572. Villemarque places the battle of Gwen-Ystrad between A.D. 547 and A.D. 560.

The British kingdom of Rheged, over which Urien ruled, is by some authorities considered as the old British or Welsh kingdom of Cumbria, or Cumberland; but, according to others, it must have been situated further northwards. In the poem of the battle of Gwen-Ystrad (see the *Myryrian Archæology*, vol. i. p. 53), Urien defeats the enemy—apparently the Saxons or Angles—under Ida, King of Bernicia. In one line near the end of the poem, Taliesin describes Urien as attacking his foes "by the white stone of Galysten:"

"Pan amwyth ai alon yn Llech wen Galysten."

The word "Galysten," when separated into such probable original components as "Gal" and "lysten," is remarkable, from the latter part of the appellation, "lysten," corresponding with the name, "Liston," of the old barony or parish in which the Cat-stane stands; the prefix Kirk (Kirk-liston) being, as is well known, a comparatively modern addition. The word "Gal" is a common term, in compound Keltic words, for "stranger," or "foreigner." In the Gaelic branch of the Keltic, "lioston" signifies, according to Sir James Foulis, "an inclosure on the side of a river." (See Mr. Muckarsie on the origin of the name of Kirkliston, in the *Statistical Account of Scotland*, vol. x. p. 68.) The Highland Society's *Gaelic Dictionary* gives "liostean" as a lodging, tent, or booth. In the Cymric, "lystyn" signifies, according to Dr. Owen Pughe, "a recess, or lodgment." (See his *Welsh Dictionary, sub voce.*) The compound word Gal-lysten would perhaps not be thus overstrained, if it were held as possibly originating in the meaning, "the lodgment, inclosure, or resting-place of the foreigner;" and the line quoted would, under such an idea, not inaptly apply to the grave-stone of such a foreign leader as Vetta. Urien's authority on such a question, states to me that he is of the same opinion as Lhwyd as to the age of the lettering in the Cat-stane legend.

To some minds it may occur as a seeming difficulty that the legend or inscription is in the Latin language, though the leader commemorated is Saxon. But this forms no kind of valid objection.

forces are described in the first line of the poem of the battle of Gwen-Ystrad, as "the men of Cattraeth, who set out with the dawn." Cattraeth is now believed by many eminent archæologists to be a locality situated at the eastern end of Antonine's wall, on the Firth of Forth—Callander, Carriden, or more probably the castle hill at Blackness, which contains various remains of ancient structures. Urien's foes at the battle of Gwen-Ystrad were apparently the Angles or Saxons of Bernicia—this last term of Bernicia, with its capital at Bamborough, including at that time the district of modern Northumberland, and probably also Berwickshire and part of the Lothians. An army marching from Cattraeth or the eastern end of Antonine's Wall, to meet such an army, would, if it took the shortest or coast line, pass, after two or three hours' march, very near the site of the Cat-stane. A ford and a fort are alluded to in the poem. The neighbouring Almond has plenty of fords; and on its banks the name of two forts or "caers" are still left—viz. Caerlowrie (Caer-l-Urien?) and Caer Almond, one directly opposite the Cat-stane, the other three miles below it. But no modern name remains near the Cat-stane to identify the name of "the fair or white strath." "Lenny"—the name of the immediately adjoining barony on the banks of the Almond, or in its "strath" or "dale"—presents insurmountable philological difficulties to its identification with Gwen; the L and G, or GW not being interchangeable. The valley of Strath-Broc (Broxburn)—the seat in the twelfth century of Freskyn of Strath-Broc, and consequently the cradle of the noble house of Sutherland—runs into the valley of the Almond about two miles above the Cat-stane. In this, as in other Welsh and Gaelic names, the word Strath is a prefix to the name of the adjoining river. In the word "Gwen-Ystrad," the word Strath is, on the contrary, in the unusual position of an affix; showing that the appellation is descriptive of the beauty or fairness of the strath which it designates. The valley or dale of the Almond, and the rich tract of fertile country stretching for miles to the south-west of the Cat-stane, certainly well merit such a desig-

The fact is, that all the early Romano-British inscriptions as yet found in Great Britain, are, as far as they have been discovered and deciphered, in Latin. And it is not more strange that a Saxon in the Lothians should be recorded in Latin, and not in Saxon or Keltic, than that the numerous Welshmen and others recorded on the early Welsh inscribed stones should be recorded in Latin and not in the Cymric tongue.

nation as "fair" or "beautiful" valley—"Gwen-Ystrad;" but we have not the slightest evidence whatever that such a name was ever applied to this tract. In his learned edition of *Les Bardes Bretons, Poemes du vie Siècle*, the Viscount Villemarque, in the note which he has appended to Taliesin's poem of the battle of Gwen-Ystrad, suggests (page 412) that this term exists in a modern form under the name of Queen's-strad, or Queen's-ferry—a locality within three miles of the Cat-stane. But it is certain that the name of Queensferry, applied to the well-known passage across the Forth, is of the far later date of Queen Margaret, the wife of Malcolm Canmore. Numerous manors and localities in the Lothians and around Kirkliston, end in the Saxon affix "ton," or town—a circumstance rendering it probable that Lis-ton had possibly a similar origin. And further, against the idea of the appellation of "the white stone of Galysten" being applicable to the Cat-stane, is the fact that it is, as I have already stated, a block of greenstone basalt; and the light tint which it presents, when viewed at a distance in strong sunlight—owing to its surface being covered with whitish lichen—is scarcely sufficient to have warranted a poet—indulging in the utmost poetical license—to have sung of it as "the white stone." After all, however, the adjective "wen," or "gwenn," as Villemarque writes it, may signify "fair" or "beautiful" when applied to the stone, just as it probably does when applied to the strath which was the seat of the battle—"Gwenn Ystrad."

Winchburgh, the name of the second largest village in the parish of Kirkliston, and a station on the Edinburgh and Glasgow Railway, is perhaps worthy of note, from its being placed in the same district as the stone of Vetta, the son of Victa, and from the appellation possibly signifying originally, according to Mr. Kemble (our highest authority in such a question), the burgh of Woden, or Wodensburgh. (See his *History of the Anglo-Saxons*, vol. i. p. 346.)

Doubtless, the Romanised Britons and the foreign colonists settled among them were, with their descendants, more or less acquainted with Latin in both its spoken and written forms. As early as the second year of his march northward for the conquest of this more distant part of Britain, or A.D. 79, Agricola, as Tacitus takes special care to inform us, took all possible means to introduce, for the purposes of conquest and civilisation, a knowledge of the Roman language and of the liberal arts among the barbarian tribes whom he went to subdue.[1] The same policy was no doubt continued to a greater or less extent during the whole era of the Roman dominion here as elsewhere; so that there is no wonder that such arts as lapidary writing, and the composition of brief Latin inscriptions, should have been known to and transmitted to the native Britons. There was, however, another class of inhabitants, besides these native Britons, who were, as we know from the altars and stone monuments which they have left, sufficiently learned in the formation and cutting of inscriptions in Latin,—a language which was then, and for some centuries subsequently, the only language used in this country, either in lapidary or other forms of writing. The military legions and cohorts which the Roman emperors employed to keep Britain under due subjection, obtained, under the usual conditions, grants of lands in the country, married, and became betimes fixed inhabitants. When speaking of the veteran soldiers of Rome settling down at last as permanent proprietors of land in Britain—as in other Roman colonies,—Sir Francis Palgrave remarks, "Upwards of forty of these barbarian legions, *some of Teutonic origin*, and others Moors, Dalmatians, and Thracians,

[1] *Vita Agricolæ*, xliv. 2.

whose forefathers had been transplanted from the remotest parts of the empire, obtained their domicile in various parts of our island, though principally upon the northern and eastern coasts, and *in the neighbourhood of the Roman walls*."[1] Such colonists undoubtedly possessed among their ranks, and were capable of transmitting to their descendants, a sufficient knowledge of the Latin tongue, and a sufficient amount of art, to form and cut such stone inscriptions as we have been considering; and perhaps I may add, that in such a mixed population, the Teutonic elements[2] in particular, would, towards the decline of the Roman dominion and power, not perhaps be averse to find and follow a leader, like Vetta, belonging to the royal stock of Woden; nor would they likely fail to pay all due respect, by the raising of a monument or otherwise, to the memory of a chief of such an illustrious race, if he fell amongst them in battle.

Besides, a brief incidental remark in Bede's History proves that the erection of a monument like the Cat-stane, to record the resting-place of the early Saxon chiefs, was not unknown. For, after telling us that Horsa was slain in battle by the Britons, Bede adds that "this Saxon leader was buried in the eastern parts of Kent, where a monument bearing his name is still in existence"[3] (hactenus in orientalibus Cantiæ partibus monumentum habet suo nomine insigne).[4] The great durability of the stone forming Vetta's

[1] *History of England*—Anglo-Saxon Period, p. 20.

[2] On the probable great extent of the Teutonic or German element of population in Great Britain as early as about A.D. 400; see Mr. Wright, in his excellent and interesting work *The Celt, the Roman, and the Saxon*, p. 385.

[3] *Historia Ecclesiastica*, lib. i. c. 1; or Dr. Giles' *Translation*, in Bohn's edition, p. 5.

[4] Dr. Giles' *Translation*, in Bohn's edition, p. 24.

monument has preserved it to the present day; while the more perishable material of which Horsa's was constructed has made it a less faithful record of that chief, though it was still in Bede's time, or in the eighth century, "suo nomine insigne."[1]

The chief points of evidence which I have attempted to adduce in favour of the idea that the Cat-stane commemorates the grandfather of Hengist and Horsa may be summed up as follows:—

1. The surname of VETTA upon the Cat-stane is the name of the grandfather of Hengist and Horsa, as given by our oldest genealogists.

2. The same historical authorities all describe Vetta as the son of Victa; and the person recorded on the Cat-stane is spoken of in the same distinctive terms—"VETTA F(ILIUS) VICTI."

3. Vetta is not a common ancient Saxon name, and it is highly improbable that there existed in ancient times two historical Vettas, the sons of two Victas.

4. Two generations before Hengist and Horsa arrived in England, a Saxon host—as told by Ammianus—was leagued with the other races of modern Scotland (the Picts, Scots, and Attacots), in fighting with a Roman army under Theodosius.

5. These Saxon allies were very probably under a leader who claimed royal descent from Woden, and consequently under an ancestor or pre-relative of Hengist and Horsa.

6. The battle-ground between the two armies was, in part at least, the district placed between the two Roman walls, and consequently included the tract in which the Cat-stane is placed; this

[1] *Historia Ecclesiastica*, lib. i. c. 15.

district being erected by Theodosius, after its subjection, into a fifth Roman province.

7. The palæographic characters of the inscription accord with the idea that it was cut about the end of the fourth century.

8. The Latin is the only language[1] known to have been used in British inscriptions and other writings in these early times by the Romanised Britons and the foreign colonists and conquerors of the island.

9. The occasional erection of monuments to Saxon leaders is proved by the fact mentioned by Bede, that in his time, or in the eighth century, there stood in Kent a monument commemorating the death of Horsa.[2]

If, then, as these reasons tend at least to render probable, the Cat-stane be the tombstone of Vetta, the grandfather of Hengist and Horsa, this venerable monolith is not only interesting as one of our most ancient national historic monuments, but it corroborates the floating accounts of the early presence of the Saxons upon our coast; it presents to us the two earliest individual Saxon names known in British history; it confirms, so far as it goes, the accur-

[1] Perhaps it is right to point out, as exceptions to this general observation, a very few Greek inscriptions to Astarte, Hercules, Esculapius, etc., left in Britain by the Roman soldiers and colonists.

[2] On the supposed site, etc. of this monument to Horsa, in Kent, see Mr. Colebrook's paper in *Archæologia*, vol. ii. p. 167; and Halsted's *Kent*, vol. ii. p. 177. In 1631, Weever, in his *Ancient Funeral Monuments*, p. 317, acknowledges that "stormes and time have devoured Horsa's monument." In 1659 Phillpot, when describing the cromlech called Kits Coty House—the alleged tomb of Catigern—speaks of Horsa's tomb as utterly extinguished "by storms and tempests under the conduct of time."

acy of the genealogy of the ancestors of Hengist and Horsa, as recorded by Bede and our early chroniclers; while at the same time it forms in itself a connecting link, as it were, between the two great invasions of our island by the Roman and Saxon—marking as it does the era of the final declinature of the Roman dominion among us, and the first dawn and commencement of that Saxon interference and sway in the affairs of Britain, which was destined to give to England a race of new kings and new inhabitants, new laws, and a new language.

ON SOME SCOTTISH MAGICAL CHARM-STONES, OR CURING-STONES.

THROUGHOUT all past time, credulity and superstition have constantly and strongly competed with the art of medicine. There is no doubt, according to Pliny, that the magical art began in Persia, that it originated in medicine, and that it insinuated itself first amongst mankind under the plausible guise of promoting health.[1] In proof of the antiquity of the belief, this great Roman encyclopædist cites Eudoxus, Aristotle, and Hermippus, as averring that magical arts were used thousands of years before the time of the Trojan war.

Assuredly, in ancient times, faith in the effects of magical charms, amulets, talismans, etc., seems to have prevailed among all those ancient races of whom history has left any adequate account. In modern times a belief in their efficiency and power is still extensively entertained amongst most of the nations of Asia and Africa. In some European kingdoms, also, as in Turkey, Italy, and Spain, belief in them still exists to a marked extent. In our own country, the magical practices and superstitions of the older and darker ages persist only as forms and varieties, so to speak, of archæological relics,—for they remain at the present day in comparatively a very

[1] *Natural History*, Book xxx. chapters i. ii.

sparse and limited degree. They are now chiefly to be found among the uneducated, and in outlying districts of the kingdom. But still, some practices, which primarily sprung up in a belief in magic, are carried on, even by the middle and higher classes of society, as diligently as they were thousands of years ago, and without their magical origin being dreamed of by those who follow them. The coral is often yet suspended as an ornament around the neck of the Scottish child, without the potent and protective magical and medicinal qualities long ago attached to it by Dioscorides and Pliny being thought of by those who place it there. Is not the egg, after being emptied of its edible contents, still, in many hands, as assiduously pierced by the spoon of the eater as if he had weighing upon his mind the strong superstition of the ancient Roman, that—if he omitted to perforate the empty shell—he incurred the risk of becoming spell-bound, etc.? Marriages seem at the present day as much dreaded in the month of May as they were in the days of Ovid, when it was a proverbial saying at Rome that

"Mense malas *Maio* nubere vulgus ait."

And, in the marriage ceremony itself, the finger-ring still holds among us as prominent a place as it did among the superstitious marriage-rites of the ancient pagan world. Among the endless magical and medical properties that were formerly supposed to be possessed by human saliva, one is almost universally credited by the Scottish schoolboy up to the present hour; for few of them ever assume the temporary character of pugilists without duly spitting into their hands ere they close their fists; as if they retained a full reliance on the magical power of the saliva to increase the strength of the impending blow—if not to avert any feeling of malice produced by it—as was enunciated, eighteen centuries ago, by one of the most laborious and esteemed writers of that age,[1] in a division of his work which he gravely prefaces with the assertion that in this special division he has made it his "object (as he declares) to state no facts but such as are established by nearly uniform testimony."

In a separate chapter (chap. iv.) in his 30th Book, Pliny alludes to the prevalence of magical beliefs and superstitious practices in the ancient Celtic provinces of France and Britain. "The Gaelic provinces," says he, "were pervaded by the magical art, and that even down to a period within memory; for it was the Emperor Tiberius who put down the Druids and all that tribe of wizards and physicians." We know, however, from the ancient history of France posterior to Pliny's time, that the Druids survived as a powerful class in that country for a long time afterwards. Writing towards the end of the first century, Pliny goes on to remark;—"At the present day, struck with fascination, Britannia still cultivates this art, and that with ceremonials so august, that she might almost seem to have been the first to communicate them to the people of Persia." "To such a degree," adds this old Roman philosopher, "are

[1] "What we are going to say," observes Pliny, "is marvellous, but it may easily be tested by experiment. If a person repents of a blow given to another, either by hand or with a missile, he has nothing to do but to spit at once into the palm of the hand which has inflicted the blow, and all feeling of resentment will be instantly alleviated in the person struck. This, too, is often verified in the case of a beast of burden, when brought on its haunches with blows: for, upon this remedy being adopted, the animal will immediately step out and mend its pace. Some persons, also, before making an effort, spit into the hand in the manner above stated, in order to make the blow *more* heavy."—Pliny's *Natural History*, xxviii. § 7.

nations throughout the whole world, totally different as they are, and quite unknown to one another, in accord upon *this* one point."[1]

Some supposed vestiges of a most interesting kind, of very ancient Gallic or Celtic word-charms, have recently been brought before archæologists by the celebrated German philologist Grimm, and by Pictet of Geneva. Marcellus, the private physician of the Roman Emperor Theodosius, was a Gaul born in Aquitane, and hence, it is believed, was intimately acquainted with the Gaulish or Celtic language of that province. He left a work on quack medicines (*De Medicamentis Empiricis*), written probably near the end of the fourth century. This work contains, amongst other things, a number of word-charms, or superstitious cure-formulas, that were, till lately, regarded—like Cato's word-cure for fractures of the bones—as mere unmeaning gibberish. Joseph Grimm and M. Pictet, however, think that they have found in these word-charms of Marcellus, specimens of the Gaulish or Celtic language several centuries older than any that were previously known to exist—none of the earliest glosses used by Zeuss, in his famous *Grammatica Celtica*, being probably earlier than the eighth or ninth centuries. If the labours of Grimm and Pictet prove successful in this curious field of labour, they will add another proof to the prevalence of magical charms

[1] *Natural History*, Book xxx. § 4. Archæologists are now fully aware of "the accord" of the ancient inhabitants of Britain with those of Persia and the other eastern branches of the Aryan race in many other particulars, as in their language, burial customs, etc. According to some Indian observers, stone erections, like our so-called Druidical circles, cromlechs, etc., are common in the East. Is it vain to hope that amid the great and yet unsearched remains of old Sanscrit literature, allusions may yet be found to such structures, that may throw more light upon their uses in connection with religious, sepulchral, or other services?

among the Celtic nations of antiquity, and afford us additional confirmation of the ancient prevalence, as described by Pliny, of a belief in the magical art among the Gaelic inhabitants of France and Britain.[1]

The long catalogue of the medical superstitions and magical practices originally pertaining to our Celtic forefathers, was no doubt from time to time increased and swelled out in Britain by the addition of the analogous medical superstitions and practices of the successive Roman [2] and Teutonic [3] invaders and conquerors of our

[1] Grimm thinks that the formulæ of Marcellus partake more of the Celtic dialects of the Irish, and consequently of the Scotch, than of the Welsh. As one of the shortest specimens of Marcellus's charm-cures, let me cite, from Pictet, the following, as given in the *Ulster Journal of Archæology*, vol. iv. p. 266 :—" Formula 12. He who shall labour under the disease of watery (or blood-shot) eyes, let him pluck the herb Millefolium up by the roots, and of it make a hoop, and look through it, saying three times, '*Excicumacriosos* ;' and let him as often move the hoop to his mouth, and spit through the middle of it, and then plant the herb again." "I divide," observes Pictet, "the formula thus : *exci cuma criosos*, and translate it, 'See the form of the girdle.'" After a long and learned disquisition on the component words Pictet adds—"The process of cure recommended in this formula is of a character altogether symbolical. Girdles (*cris*), which we shall meet with again in formula No. 27, seem to have performed an important part in Celtic medicine. By making the eye look through the circle formed by the plant, a girdle, as it were, was put round it ; and it is for this reason that the formula says, See the form (or model) of the girdle. The action of spitting afterwards through the little ring expressed symbolically the expulsion of the pain." The so-called Celtic word-charms in the formulæ of Marcellus are usually longer than the above ; as, "*Tetune resonco bregan gresso* ;" "Heilen prossaggeri nome sipolla na builet ododieni iden olitan," etc. etc.

[2] On this subject I elsewhere published, two years ago, the following remarks :—" The medical science and medical lore of the past has become,

[3] See, for example, Kemble's work on the Anglo-Saxons, vol. i. p. 528, for various Teutonic medical superstitions and cures.

island. A careful analysis would yet perhaps enable the archæologist to separate some of these classes of magical beliefs from each other; but many of them had, perhaps, a common and long anterior origin. We know further that, in its earlier centuries among us, the teachers of Christianity added greatly to the number of existing medical superstitions, by maintaining the efficacy, for example, of a visit to the cross of King Edwin of Northumberland, for the cure of agues, etc.,—the marvellous alleged recoveries worked by visiting the grave of St. Ninian at Whitehorn, or the cross of St. Mungo in the Cathedral churchyard at Glasgow; the sovereign virtues of the waters of wells used by various anchorets, and dedicated to various

after a succession of ages, the so-called folk-lore and superstitious usages of times nearer our own. Up to the end of the last century, patients attacked with insanity were occasionally dipped in lakes and wells, and left bound in the neighbouring church for a night. Loch Maree, in Ross-shire, and St. Fillan's Pool, in Perthshire, were places in which such unfortunate patients were frequently dipped. Heron, in his *Journey through Scotland* in the last century, states that it was affirmed that two hundred invalids were carried annually to St. Fillan's for the cure of various diseases, but principally of insanity. The proceedings at this famous pool were in such cases an imitation of the old Greek and Roman worship of Æsculapius. Patients consulting the Æsculapian priest were purified first of all, by bathing in some sacred well; and then having been allowed to enter into and sleep in his temple, the god, or rather some priest of the god, came in the darkness of the night and told them what treatment they were to adopt. The poor lunatics brought to St. Fillan's were, in the same way, first purified by being bathed in his pool, and then laid bound in the neighbouring church during the subsequent night. If they were found loose in the morning, a full recovery was confidently looked for, but the cure remained doubtful when they were found at morning dawn still bound. I was lately informed by the Rev. Mr Stewart of Killin, that in one of the last cases so treated—and that only a few years ago—the patient was found sane in the morning, and unbound; a dead relative, according to the patient's own account, having entered the church during the

saints throughout the country; the curative powers of holy robes, bells, bones, relics, etc.

Numerous forms of medical superstitions, charms, amulets, incantations, etc., derived from the preceding channels, and possibly also from other sources, seem to have been known and practised among our forefathers, and for the cure of almost all varieties of human maladies, whether of the mind or body. Our old Scottish hagiologies, witch trials, ecclesiastical records, etc., abound with notices of them. Nor have some of the oldest and most marked medical superstitions of ancient times been very long obliterated and forgotten. I know, for example, of two localities in the Lowlands, one near Biggar in Lanarkshire, the other near Torphichen in West Lothian, where, within the memory of the present and past

night, and loosened her both from the ropes that bound her body and the delusions that warped her mind. It was a system of treatment by mystery and terrorism that might have made some sane persons insane; and hence, perhaps, conversely, some insane persons sane. Mr. Pennant tells us that at Llandegla, in Wales, where similar rites were performed for the cure of insanity, viz., purification in the sacred well, and forced detention of the patient for a night in the church, under the communion-table, the lunatics or their friends were obliged to leave a cock in the church if he were a male, and a hen if she were a female—an additional circumstance in proof of the Æsculapian type of the superstition. But perhaps, after all, the whole is a medical or mythological belief, older than Greece or Rome, and which was common to the whole Aryan or Indo-European race in Asia before they sent off, westward, over Europe, those successive waves of population that formed the nations of the Celt and Teuton, of the Goth, and Greek, and Latin. The cock is still occasionally sacrificed in the Highlands for the cure of epilepsy and convulsions. A patient of mine found one, a few years ago, deposited in a hole in the kitchen floor; the animal having been killed and laid down at the spot where a child had, two or three days previously, fallen down in a fit of convulsions." —See the *Medical Times and Gazette* of Dec. 8, 1860, p. 549.

generation, living cows have been sacrificed for curative purposes, or under the hope of arresting the progress of the murrain in other members of the flock. In both these instances the cow was sacrificed by being buried alive. The sacrifice of other living animals,[1] as of the cat, cock, mole, etc., for the cure of disease, and especially of fits, epilepsy, and insanity, continues to be occasionally practised in some parts of the Highlands up to the present day. And in the city of Edinburgh itself, every physician knows the fact that, in the chamber of death, usually the face of the mirror is most carefully covered over, and often a plate with salt in it is placed upon the chest of the corpse.

[1] A very intelligent patient from the North Highlands, to whom I happened lately to speak on this subject, has written out the following instances that have occurred within her own knowledge :—"Twenty years or more ago, in the parish of Nigg, Ross-shire, there was a lad of fifteen ill with epilepsy. To cure him, his friends first tried the charm of mole's blood. A plate was laid on the lad's head ; the living mole was held over it by the tail, the head cut off, and the blood allowed to drop into the plate. Three moles were sacrificed one after the other, but without effect. Next they tried the effect of a bit of the skull of a suicide, and sent for this treasure a distance of from sixty to one hundred miles. This bit of the skull was scraped to dust into a cup of water, which the lad had to swallow, not knowing the contents. This I heard from a sister of the lad's. There was a 'strong-minded' old woman at Strathpeffer, Ross-shire whose daughter told me that the neighbours had come to condole with the mother after she had fallen down in a fit of some kind. They strongly advised her to bury a living cock in the very place where she had fallen, to prevent a return of the ailment. A woman in Sutherlandshire told me that she knew a young man, ill of consumption, who was made to drink his own blood after it had been drawn from his arm. This same woman was ill with a pain in her chest, which she could get nothing to relieve ; so her father sent off for 'a knowing man,' who, when he saw the girl, repeated some words under his breath, then touched the floor and her shoulder three times alternately, and with alleged success."

The Museum of the Society contains a few medicinal charms and amulets, principally in the form of amber beads (which were held potent in the cure of blindness), perforated stones, and old distaff whorls, whose original use seems to have been forgotten, and new and magical properties assigned to them. But the most important medicinal relic in the collection is the famous "Barbreck's bone," a slice or tablet of ivory, about seven inches long, four broad, and half-an-inch in thickness. It was long in the possession of the ancient family of Barbreck in Argyleshire, and over the Western Highlands had the reputation of curing all forms and degrees of insanity. It was formerly reckoned so valuable that a bond of £100 was required to be deposited for the loan of it.

But the main object of the present communication is, through the kind permission of Struan Robertson, Lady Lockhart of Lee, and others, to show to the Society two or three of the principal curing-stones of Scotland.

Several of these curing-stones long retained their notoriety, but they have now almost all fallen entirely into disuse, at least for the cure of human diseases. In some districts, however, they are still employed in the treatment of the diseases of domestic animals.

A very ancient example of the use of a "curing-stone," in this country is detailed in what may be regarded as the first or oldest historical work which has been left us in reference to Scotland, namely, in Adamnan's *Life of St. Columba.* This biography of the founder of Iona was probably written in the last years of the seventh century, Adamnan having died in A.D. 705. He was elected to the Abbacy of Iona A.D. 679, and had there the most favourable opportunities of becoming acquainted with all the existing traditions and records regarding St. Columba. About the year 563 of the Christian era, Columba visited Brude, King of the Picts, in his royal fort on the Ness, and found the Pictish sovereign attended by a court or council, and with Brochan as his chief Druid or Magus. Brochan retained an Irish female, and consequently a countrywoman of Columba's, as a slave. The 33d chapter of the second book of Adamnan's work is entitled, "Concerning the Illness with which the Druid (*Magus*) Brochan was visited for refusing to liberate a Female Captive, and his Cure when he restored her to Liberty." The story told by Adamnan, under this head, is as follows :—

Curing-Stone of St. Columba.

"About the same time the venerable man, from motives of humanity, besought Brochan the Druid to liberate a certain Irish female captive, a request which Brochan harshly and obstinately refused to grant. The Saint then spoke to him as follows :—'Know, O Brochan, know, that if you refuse to set this captive free, as I desire you, you shall die before I return from this province.' Having said this in presence of Brude the king, he departed from the royal palace and proceeded to the river Nesa, from which he took a white pebble, and showing it to his companions, said to them :—'Behold this white pebble, by which God will effect the cure of many diseases.' Having thus spoken, he added, 'Brochan is punished grievously at this moment, for an angel sent from heaven, striking him severely, has broken in pieces the glass cup which he held in his hands, and from which he was in the act of drinking, and he himself is left half dead. Let us await here, for a short time, two of the king's messengers, who have been sent after us in haste, to request us to return quickly and relieve the dying Brochan, who, now that he is thus terribly punished, consents to set his captive free.'

"While the saint was yet speaking, behold, there arrived as he had predicted, two horsemen, who were sent by the king, and who related all that had occurred, according to the prediction of the saint—the breaking of the drinking goblet, the punishment of the Druid, and his willingness to set his captive at liberty. They then added :—'The king and his councillors have sent us to you to request that you would cure his foster father, Brochan, who lies in a dying state.'

"Having heard these words of the messengers, Saint Columba sent two of his companions to the king, with the pebble which he had blessed, and said to them ; 'If Brochan shall first promise to free his captive, immerse this little stone in water and let him drink from it, but if he refuse to liberate her, he will that instant die.'

"The two persons sent by the saint proceeded to the palace and announced the words of the holy man to the king and to Brochan, an announcement which filled them with such fear, that he immediately liberated the captive and delivered her to the saint's messengers."

The stone was then immersed in water, and in a wonderful manner, and contrary to the laws of nature, it floated on the water like a nut or an apple, nor could it be submerged. Brochan drank from the stone as it floated on the water, and instantly recovered his perfect health and soundness of body.

"This little pebble (adds Adamnan) was afterwards preserved among the treasures of the king, retained its miraculous property of floating in water, and through the mercy of God effected the cure of sundry diseases. And, what is very wonderful, when it was sought for by those sick persons whose term of life had arrived it could not be found. An instance of this occurred the very day king Brude died, when the stone, though sought for with great diligence, could not be found in the place where it had been previously left."[1]

In the Highlands of Scotland there have been transmitted down, for many generations, various curing or charm-stones, used in the same manner as that of Columba, and reckoned capable, like his, of imparting to the *water in which they were immersed*[2] wondrous medicinal powers. One of the most celebrated of these curing-stones belongs to Struan Robertson, the chief of the Clan Donnachie. I am indebted to the kindness of Mrs. Robertson, for the following notes regarding the curing-stone, of which her family are the hereditary proprietors. Its local name is

Clach-na-Bratach, or Stone of the Standard.

"This stone has been in possession of the Chiefs of Clan Donnachaidh since 1315.

"It is said to have been acquired in this wise.

[1] In the first chapter of Adamnan's work, the miracle is again alluded to as follows:—"He took a white stone (*lapidem candidum*) from the river's bed, and blessed it for the cure of certain diseases; and that stone, contrary to the law of nature, floats like an apple when placed in the water."

[2] For other instances of waters rendered medicinal by being brought in contact with saint's bones—such as St. Marnan's head, with St. Conval's chariot, etc. etc., see Dalyell's *Superstitions of Scotland*, p. 151, etc. Sibbald's *Memoirs of the Edinburgh College of Physicians*, p. 39.

"The (then) chief, journeying with his clan to join Bruce's army before Bannockburn, observed, on his standard being lifted one morning, a glittering something in a clod of earth hanging to the flag-staff. It was this stone. He showed it to his followers, and told them he felt sure its brilliant lights were a good omen and foretold a victory—and victory was won on the hard-fought field of Bannockburn.

Fig. 17. Clach-na-Bratach.

"From this time, whenever the clan was 'out,' the Clach-na-Bratach accompanied it, carried on the person of the chief, and its varying hues were consulted by him as to the fate of battle. On the eve of Sheriffmuir (13th November 1715), of sad memory, on Struan consulting the stone as to the fate of the morrow, the large internal flaw was first observed. The Stuarts were lost—and Clan Donnachaidh has been declining in influence ever since.

"The virtues of the Clach-na-Bratach are not altogether of a martial nature, for it cures all manner of diseases in cattle and horses, and formerly in human beings also, if they drink the water in which this charmed stone has been thrice dipped by the hands of Struan."

The Clach-na-Bratach is a transparent, globular mass of rock crystal, of the size of a small apple. (See accompanying woodcut, Fig. 17.) Its surface has been artificially polished. Several specimens of round rock-crystal, of the same description and size, and similarly polished, have been found deposited in ancient sepulchres, and were formerly used also in the decoration of shrines and sceptres.

Another well-known example of the Highland curing-stone is the

Clach Dearg, or Stone of Ardvoirloch.

This stone is a clear rock-crystal ball of a similar character, but somewhat smaller than the Clach-na-Bratach, and placed in a setting (see Fig. 18) of four silver bands or slips. The following account of the Ardvoirloch curing-stone is from the pen of one of the present members of that ancient family:—

Fig. 18. Stone of Ardvoirloch.

"It has been in the possession of our family from *time immemorial*, but there is no writing about it in any of the charters, nor even a tradition as to *when* and *how* it became possessed of it. It is supposed to have been brought from the *East*, which supposition is corroborated by the fact of the silver setting being recognised as of Eastern workmanship. Its healing powers have always been held in great repute in our own neighbourhood, particularly in diseases of cattle. I have even known persons come for the water into which it has been dipped from a distance of forty miles. It is also believed to have other properties which you know of.

"These superstitions would have existed up to the present day, had I not myself put a stop to them; but six years ago, I took an opportunity to do away with them, by depositing the stone with some of the family plate in a chest which I sent to the bank. Thus, when applied to for it (which I have been since then), I had the excuse of not having it in my possession; and when the Laird returns from India, it is hoped the superstition may be forgotten, and "the stone" preserved only as a very precious *heirloom.*

"I may mention that there were various forms to be observed by those who wished to benefit by its healing powers. The person who came for it to Ardvoirloch was obliged to draw the water himself, and bring it into the house in some vessel into which this stone was to be dipped. A bottle was filled and carried away; and in its conveyance home, if carried into any house by the way, the virtue was supposed to leave the water; it was therefore necessary, if a visit had to be paid, that the bottle should be left outside."

Other charm-stones enjoyed, up to the present century, no small medical reputation among the inhabitants of the Highlands. In some districts, every ancient family of note appears to have affected the possession of a curing-stone. The Campbells of Glenlyon have long been the hereditary proprietors of a charm-stone similar to those that I have already mentioned. It consists of a roundish or ovoidal ball, apparently of rock-crystal, about an inch and a half in diameter, and protected by a silver mounting. To make the water in which it was dipped sufficiently medicinal and effective, the stone, during the process, required to be held in the hand of the Laird. The Bairds of Auchmeddan possessed another of these celebrated northern amulets. The Auchmeddan Stone is a ball of black-coloured flint, mounted with four strips of silver. A legend engraved on this silver setting—in letters probably of the last century—states that this "Amulet or charm belonged to the family of Baird of Auchmeddan from the year 1174." In the middle of the last century, this amulet passed as a family relic to the Frasers of Findrack, when an intermarriage with the Bairds occurred.

Curing-stones seem to have formerly been by no means rare in this country, to the south also of the Highland Borders. In a letter written by the distinguished Welsh archæologist Edward Llwyd, and dated Linlithgow, December 17, 1699, he states that betwixt Wales and the Highlands he had seen at least fifty different forms of the party-coloured glass bead or amulet known under the name of Adder-beads or Snake stones.[1] In Scotland he found various materials used as healing amulets, particularly some pebbles of remarkable shape and colour, and hollow balls and rings of coloured glass. "They have also," he says, "the *Ombriæ pellucidæ*, which are crystal balls or hemispheres, or depressed ovals, in great esteem for curing of cattle; and some on May-day put them into a tub of water, and besprinkle all their cattle with that water, to prevent being elf-struck, bewitched, etc."

In the Lowlands, the curing-stone of greatest celebrity, and the one which has longest retained its repute, is

The Lee Penny.

In the present century this ancient medical charm-stone has acquired a world-wide reputation as the original of the *Talisman* of Sir Walter Scott, though latterly its therapeutic reputation has

[1] See *Philosophical Transactions* for the year 1713, p. 98. For instances of curing-stones in the Hebrides, see Martin's *Western Isles*, p. 134, 166, etc.

greatly declined, and almost entirely ceased.[1] The enchanted stone has long been in the possession of the knightly family of the Lockharts of Lee, in Lanarkshire. According to a mythical tradition, it was, in the fourteenth century, brought by Sir Simon Lockhart from the Holy Land, where it had been used as a medical amulet, for the arrestment of hæmorrhage, fever, etc. It is a small dark-red stone, of a somewhat triangular or heart shape, as represented in the adjoining woodcut (Fig. 19). It is set in the reverse of a groat of Edward IV., of the London Mint.[2]

Fig. 19. The Lee Penny.

When the Lee Penny was used for healing purposes, a vessel was filled with water, the stone was drawn once round the vessel, and then dipped three times in the water. In his *Account of the Penny in the Lee*, written in 1702, Hunter states, that "it being taken and put into the end of a cloven stick, and washen in a tub full of

[1] I was lately told by the farmer at Nemphlar, in the neighbourhood of Lee, that in his younger days no byre was considered safe which had not a bottle of water from the Lee Penny suspended from its rafters. Even this remnant of superstition seems to have died out during the present generation

[2] I state this on the high numismatic authority of my friend, Mr. Sim. Sir Walter Scott describes the coin as a groat of Edward I.

water, and given to cattell to drink, infallibly cures almost all manner of deseases. The people," he adds, "come from all airts of the kingdom with deseased beasts."

One or two points in its history prove the faith that was placed in the healing powers of the Lee Penny in human maladies of the most formidable type. About the beginning of last century, Lady Baird of Saughtonhall was attacked with the supposed symptoms of hydrophobia. But on drinking of, and bathing in, the water in which the Lee Penny had been dipped, the symptoms disappeared; and the Knight and Lady of Lee were for many days sumptuously entertained by the grateful patient. In one of the epidemics of plague which attacked Newcastle in the reign of Charles I., the inhabitants of that town obtained the loan of the Lee Penny by granting a bond of £6000 for its safe return. Such, it is averred, was their belief in its virtues, and the good that it effected, that they offered to forfeit the money, and keep the charm-stone.

About the middle of the seventeenth century the Reformed Protestant Church of Scotland zealously endeavoured, as the English Church under King Edgar had long before done, to "extinguish every heathenism, and forbid well-worshippings, and necromancies, and divinations, and enchantments, and man-worshippings, and the vain practices which are carried on with various spells, and with elders, and also with other trees, and with stones, etc."[1] They left, however, other practices, equally superstitious, quite untouched. Thus, while they threatened "the seventh son of a woman" with the "paine of Kirk censure," for "cureing the cruelles (scrofulous

[1] Kemble's *Anglo-Saxons*, vol. i. p. 527, etc.

tumours and ulcers),"[1] by touching them, they still allowed the reigning king this power (Charles II. alone "touched" 92,000 such patients);[2] and the English Church sanctioned a liturgy to be used on these superstitious occasions. Again, the Synod of the Presbyterian Church of Glasgow examined into the alleged curative gifts of the Lee Penny; but, finding that it was employed "wtout using onie words such as charmers and sorcerers use in their unlawfull practisess; and considering that in nature there are mony things seen to work strange effects, q[r]of no human witt can give a reason, it having pleasit God to give to stones and herbes special virtues for the healing of mony infirmities in man and beast, advises the brethern to surcease their process, as q[r]in they perceive no ground of offence: And admonishes the said Laird of Lee, in the useing of the said stone to tak heed that it be used hereafter w[t] the least scandal that possiblie may be."[3]

[1] See a case of this prohibition in the *Ecclesiastical Records of the Presbytery of St. Andrews* for September 1643. "It is manifest by experience," says Upton, "that the seventh male child by just order, never a girle or wench being borne betweene, doth heall only with touching, by a natural gift, the king's evil; which is a speciall gift of God, given to kings and queens, as daily experience doth witnesse." See Upton's Notable Things (1631), p. 28. Charles I. when he visited Scotland in 1633, in Holyrood Chapel, on St. John's day, "heallit 100 persons of the cruelles, or kingis eivell, yong and olde."—Dalyell's *Superstitions*, p. 62.

[2] See the "*Charisma Basilicon*" (1684) of John Browne, "Chirurgion to His Majesty," for a full and charming account of the whole process and ceremonies of the royal "touch," the prayers used on the occasion, and due proofs of the alleged wondrous effects of this "sanative gift, which hath (says Dr. Browne) for above 640 years been confirmed and continued in our English Princely line, wherein is not so much of their Majesty shown as of their Divinity," and which is only doubted by "Ill affected men and Dissenters."

[3] See the *Gentleman's Magazine* for December 1787.

IS THE GREAT PYRAMID OF GIZEH A METROLOGICAL MONUMENT?

The following observations form a corrected Abstract, from No. 75 of the *Proceedings of the Royal Society of Edinburgh*, of a communication made to that Society on the 20th January 1868, and entitled *Pyramidal Structures in Egypt and elsewhere; and the Objects of their Erection*. Some additional points are dwelt upon in the Notes and Appendix. As stated at the time, the communication was not at all spontaneous, but enforced by the previous criticisms of Professor Smyth.

THERE are many proposed derivations of the word Pyramid. Perhaps the origin of the name suggested by the distinguished Egyptologist, Mr. Birch, from two Coptic words, "*pouro*," "the king," and "*cmahau*," or "*maha*," "tomb,"—the two in combination signifying "the king's tomb,"—is the most correct. "*Men*," in Coptic, signifies "monument," "memorial;" and "*pouro-men*," or "king's monument," may possibly also be the original form of the word.[1]

Various English authors, as Pope,[2] Pownall,[3] Professor Daniel

[1] See on other proposed significations and origins of the word pyramid, APPENDIX, No. I.

[2] In the plain of Troy, and on the higher grounds around it, various barrows still remain, and have been described from Pliny, Strabo, and Lucian

[3] Colonel Pownall, while describing in 1770 the barrow of New Grange, in Ireland, to the London Society of Antiquaries, speaks of it as "a pyramid of stone." "This pyramid," he observes, "was encircled at its base with a number of enormous unhewn stones," etc. "The pyramid, in its present state, is but a ruin of what it was," etc. etc. See *Archæologia*, vol. vi. p. 254; and Higgins' *Celtic Druids*, p. 40, etc.

Wilson,[1] Burton,[2] had long applied the term pyramid to the larger forms of conical and round sepulchral mounds, cairns, or barrows—such as are found in Ireland, Brittany, Orkney, etc., and also in numerous districts of the New and Old World;[3] and which are all

down to Lechevalier, Forchhammer, and Maclaren. In later times, Choiseul and Calvert have opened some of them. Homer gives a minute account of the obsequies of Patroclus and the raising of his burial-mound, which forms, as is generally believed, one of those twin barrows still existing on the sides of the Sigean promontory, that pass under the name of the tumuli of Achilles and Patroclus. Pope, in translating the passage describing the commencement of the funeral pyre, uses the word pyramid. For

. . "those deputed to inter the slain,
Heap with a rising *pyramid* the plain."

Professor Daniel Wilson, in alluding, in his *Prehistoric Annals*, vol. i. p. 74, to this account by Homer of the ancient funeral-rites, and raising of the funeral-mound, speaks of the erection of Patroclus' barrow as "the methodic construction of the Pyramid of earth which covered the sacred deposit and preserved the memory of the honoured dead."

[1] In his *Prehistoric Annals of Scotland*, Dr. Daniel Wilson states (vol. i. p. 87), that "the Chambered Cairn properly possesses as its peculiar characteristic the enclosed catacombs and galleries of megalithic masonry, branching off into various chambers symmetrically arranged, and frequently exhibiting traces of constructive skill, such as realise in some degree the idea of the regular pyramid." He speaks again of the stone barrows or cairns of Scotland as "monumental pyramids" (vol. i. p. 67); of the earth barrow being an "earth pyramid or tumulus" (p. 70); of Silbury Hill as an "earth pyramid" (p. 62): and in the same page, in alluding to the large barrow-tomb of the ancient British chief or warrior, he states, "in its later circular forms we see the rude type of the great pyramids of Egypt." The same learned author, in his work on *Prehistoric Man*, refers to the great monuments of the American mound-builders as "earth pyramids" (p. 202), "huge earth pyramids" (p. 205), "pyramidal earth-works" (p. 203); etc.

[2] In his *History of Scotland*, Mr. Burton speaks of the barrows of New Grange and Maeshowe (Orkney), as erections which "may justly be called minor pyramids" (vol. i. p. 114).

[3] In mentioning the great numbers of sepulchral barrows spread over the

characterised by containing in their interior chambers or cells, constructed usually of large stones, and with megalithic galleries leading into them. In these chambers (small in relation to the hills of stone or earth in which they were imbedded) were found the remains of sepulture, with stone weapons, ornaments, etc. The galleries and chambers were roofed, sometimes with flags laid quite flat, or placed abutting against each other; and occasionally with large stones arranged over the internal cells in the form of a horizontal arch or dome. In his travels to Madeira and the Mediterranean (1840), Sir W. Wilde details in interesting terms his visit to the pyramids of Egypt; and in describing the roof of the interior chambers of one of the pyramids at Sakkara,[1] he remarks on the analogy of its construction to the great barrow of Dowth in Ireland; and again, when writing—in his work on the *Beauties of the Boyne* (1849)—an account of the great old barrows of Dowth, New Grange,

world, Sir John Lubbock observes—"In our own island they may be seen on almost every down; in the Orkneys alone it is estimated that two thousand still remain; and in Denmark they are even more abundant; they are found all over Europe from the shores of the Atlantic to the Oural Mountains; in Asia they are scattered over the great steppes from the borders of Russia to the Pacific Ocean, and from the plains of Siberia to those of Hindostan; in America we are told that they are numbered by thousands and tens of thousands; nor are they wanting in Africa, where the pyramids themselves exhibit the most magnificent development of the same idea; so that the whole world is studded with these burial-places of the dead."—*Prehistoric Times*, p. 85. See similar remarks in Dr. Clarke's *Travels*, 4th edition, vol. i. p. 276, vol. ii. p. 75, etc.

[1] Sir J. Gardner Wilkinson thinks that the pyramids of Sakkara are probably older than the other groups of these structures, as those of Gizeh or the Great Pyramid erected during the fourth dynasty of kings.—See Rawlinson's *Herodotus*, vol. ii. chap. viii. Manetho assigns to Uenophes, one of the monarchs in the first dynasty, the erection of the Pyramids of Cochome. See Kenrick's *Ancient Egypt*, ii. p. 112, 122, 123; Bunsen's *Egypt*, ii. 99, etc.

etc., placed on its banks above Drogheda, he describes at some length the last of these mounds (New Grange),—stating that it "consists" of an enormous cairn or " hill of small stones, calculated at 180,000 tons weight, occupying the summit of one of the natural undulating slopes which enclose the valley of the Boyne upon the north. It is said to cover nearly two acres, and is 400 paces in circumference, and now about 80 feet higher than the adjoining natural surface. Various excavations (he adds) made into its sides and upon its summit, at different times, in order to supply materials for building and road-making, having assisted to lessen its original height, and also to destroy the beauty of its outline." Like the other analogous mounds and pyramids placed there and elsewhere, New Grange has a long megalithic gallery, of above 60 feet in length, leading inward into three dome-shaped chambers or crypts. After describing minutely, and with a master-hand, the construction of these interior parts, and the carvings of circles, spirals, etc.,[1] upon the enormous stones of which the gallery and crypts are built, Sir William Wilde goes on to observe :—" We believe with most modern investigators into such subjects, that it was a tomb, or great sepulchral Pyramid, similar in every respect to those now standing by the banks of the Nile, from Dashour to Gizeh, each consisting of a great central chamber containing one or more sarcophagi, entered by a long stone-covered passage. The external aperture was concealed, and the whole covered with a great mound of stones or earth in a conical form. The early Egyptians, and the

[1] On these Archaic forms of sculpture, see APPENDIX, No. II. In many barrows the gallery in its course—and in some as it enters the crypt—is contracted, and more or less occluded by obstructions of stone, etc., which Mr. Kenrick likens to the granite portcullises in the Great Pyramid. See his *Ancient Egypt*, vol. i. p. 121.

Mexicans also, possessing greater art and better tools than the primitive Irish, carved, smoothed, and cemented their great pyramids; *but the type and purpose is all the same*. . . . How far anterior to the Christian era its date should be placed would be a matter of speculation; it may be of an age coeval, or even anterior, to its brethren on the Nile."

Other pyramidal barrows at Maeshowe, Gavr Inis, etc., were referred to and illustrated; showing that a gigantic sepulchral cairn was in its mass an unbuilt pyramid; or, in other words, that a pyramid was a built cairn.

Sepulchral Character, etc., of the Egyptian Pyramids.

All authors, from the Father of History downwards, have generally agreed in considering the pyramids of Egypt as magnificent and regal sepulchres; and the sarcophagi, etc., of the dead have been found in them when first opened for the purposes of plunder or curiosity. The pyramidal sepulchral mounds on the banks of the Boyne were opened and rifled in the ninth century by the invading Dane, as told in different old Irish annals; and the Pyramids of Gizeh, etc., were reputedly broken into and harried in the same century by the Arabian Caliph, Al Mamoon,—the entrances and galleries blocked up by stones being forced and turned, and in some parts the solid masonry perforated. The largest of the Pyramids of Gizeh—or "the Great Pyramid," as it is generally termed—is now totally deprived of the external polished limestone coating which covered it at the time of Herodotus's visit, some twenty-two centuries ago; and "now" (writes Mr. Smyth) "is so injured as to be, in the eyes of some passing travellers, little better than a heap of stones." But all the internal built core of the magnificent structure remains, and contains in its interior (besides a rock chamber below) two higher built chambers or crypts above—the so-called King's Chamber and Queen's chamber—with galleries and apartments leading to them. The walls of these galleries and upper chambers are built with granite and limestone masonry of a highly-finished character. This, the largest and most gigantic of the many pyramids of Egypt, had been calculated by Major Forlong (Asso. Inst. C. Engrs.), as a structure which in the East would cost about £1,000,000. Over India, and the East generally, enormous sums had often been expended on royal sepulchres. The Taj Mahal of Agra, built by the Emperor Shah Jahan for his favourite queen, cost perhaps double or triple this sum; and yet it formed only a portion of an intended larger mausoleum which he expected to rear for himself. The great Pyramid contains in its interior, and directly over the King's Chamber, five entresols or "chambers of construction," as they have been termed, intended apparently to take off the enormous weight of masonry from the cross stones forming the roof of the King's Chamber itself. These entresols are chambers, small and unpolished, and never intended to be opened. But in two or three of them, broken into by Colonel H. Vyse, a most interesting discovery was made about thirty years ago. The surfaces of some of the stones were found painted over in red ochre or paint, with rudish hieroglyphics—being, as first shown by Mr. Birch, quarry marks, written on the stones 4000 years ago, and hence, perhaps, forming the oldest preserved writing in the world. These accidental hieroglyphics usually marked only the number and position of the individual stones. Among them, however, Mr. Birch discovered two royal ovals, viz., Shufu (the Cheops of Herodotus) and Nu Shufu—"a brother" (writes Professor Symth) "of Shufu, also a king and a co-regent with him." Most, if not all, of the other pyramids are believed to have been erected by individual kings during their individual or separate reigns. If these hieroglyphics proved that *two* kings were connected with the building of the Great Pyramid, that circumstance would perhaps account for its size and the duplicity and position of its sepulchral chambers.[1]

The pyramid standing next the Great Pyramid, and nearly of equal size, is said by Herodotus to have been raised by the brother of Cheops. The other pyramids at Gizeh are usually regarded as later in date. But the exact era of the reign or reigns of their builders has not as yet been determined, in consequence of the

[1] Mr. Birch, however—and it is impossible to cite a higher authority in such a question—holds the cartouches of Shufu and Nu Shufu to refer only to one personage—namely, the Cheops of Herodotus; and, believing with Mr. Wilde and Professor Lepsius, that the pyramids were as royal sepulchres built and methodically extended and enlarged as the reigns of their intended occupants lengthened out, he ascribes the unusual size of the Great Pyramid to the unusual length—as testified by Manetho, etc.—of the reign of Cheops; the erection of a sepulchral chamber in its built portion above being, perhaps, a step adopted in consequence of some ascertained deficiency in the rock chamber or gallery below. Indeed, the subterranean chamber under the Great Pyramid has, to use Professor Smyth's words, only been "begun to be cut out of the rock from the ceiling downwards, and left in that *unfinished* state." (Vol. i. 156.) Mr. Perring, who—as engineer—measured, worked, and excavated so very much at the Pyramids of Gizeh, under Colonel Howard Vyse, held; at the end of his researches, that "the principal chamber" in the Second Pyramid is still undetected. See Vyse's *Pyramid of Gizeh*, vol. i. 99.

break made in Egyptian chronology by the invasion of the Shepherd Kings.

In their mode of building, the various pyramids of Gizeh, etc., are all similar, and the Great Pyramid does not specially differ from the others. "There is nothing" (observes Professor Smyth) "in the stone-upon-stone composition of the Great Pyramid which speaks of the mere building problem to be solved there, as being of a different character, or requiring inventions by man of any absolutely higher order than elsewhere." But the Great Pyramid has been imagined to contain some hidden symbols and meanings. For "it is the manner of the Pyramid" (according to Professor Smyth) "not to wear its most vital truths in prominent outside positions."

Alleged Metrological Object of the Great Pyramid.

By several authorities the largest[1] of the group of pyramids at Gizeh, or "Great Pyramid," has been maintained—and particularly of late by Gabb, Jomard, Taylor, and Professor Smyth—not to be a royal mausoleum, but to be a marvellous metrological monument, built some forty centuries ago, as "a necessarily material centre," to hold and contain within it, and in its structure, material standards, "in a practicable and reliable shape," "down to the ends of the world," as measures of length, capacity, weight, etc., for men and nations for all time—"a monument" (in the language of Professor Smyth) "devoted to weights and measures, not so much as a place

[1] The Mexican Pyramid of Cholula has a base of more than 1420 feet, and is hence about twice the length of the basis of the Great Pyramid of Gizeh. See Prescott's *Conquest of Mexico*, book iii. chap. i., and book v. chap. iv.

of frequent reference for them, but one where the original standards were to be preserved for some thousands of years, safe from the vicissitudes of empires and the decay of nations." Messrs. Taylor and Smyth further hold that this Great Pyramid was built for these purposes of mensuration under Divine inspiration—the standards being, through superhuman origination and guidance, made and protected by it till they came to be understood and interpreted in these latter times. For, observes Professor Smyth, "the Great Pyramid was a sealed book to all the world *until* this present day, when modern science, aided in part by the dilapidation of the building and the structural features thereby opened up—has at length been able to assign the chief interpretations." Professor Smyth has, in his remarkable devotedness and enthusiasm, lately measured most of the principal points in the Great Pyramid; and for the great zeal, labour, and ability which he has displayed in this self-imposed mission, the Society have very properly and justly bestowed upon him the Keith Medal. But the exactitude of the measures does not necessarily imply exactitude in the reasoning upon them; and on what grounds can it be possibly regarded as a metrological monument and not a sepulchre, is legitimately the subject of our present inquiry. In such an investigation springs up first this question—

Who was the Architect of the Great Pyramid?

Mr. Taylor ascribes to Noah the original idea of the metrological structure of the Great Pyramid. "To Noah" (observes Mr. Taylor) "we must ascribe the original idea, the presiding mind, and the benevolent purpose. He who built the Ark, was of all men the most competent to direct the building of the Great Pyramid. He was born 600 years before the Flood and lived 350 years after that event, dying in the year 1998 B.C. Supposing the pyramids were commenced in 2160 B.C. (that is 4000 years ago), *they* were founded 168 years before the death of Noah. We are told" (Mr. Taylor continues) "that Noah was a 'preacher of righteousness,' but nothing could more illustrate this character of a preacher of righteousness after the Flood than that he should be the first to publish a system of weights and measures for the use of all mankind, based upon the measure of the earth." Professor Smyth, computing by another chronology, rejects the presence of Noah, and makes a shepherd—Philition, slightly and incidentally alluded to in a single passage by Herodotus [1]—the presiding and directing genius of the structure; —holding him to be a Cushite skilled in building, and under whose inspired direction the pyramid rose, containing within it miraculous measures and standards of capacity, weight, length, heat, etc.

The Coffer in the King's Chamber in the Great Pyramid an Alleged Standard for Measures of Capacity.

A granite coffer, stone box, or sarcophagus standing in that interior cell of the pyramid, called the King's Chamber, is held by

[1] Herodotus states that the Egyptians detested the memories of the kings who built the two larger Pyramids, viz., Cheops and Cephren; and hence, he adds, they commonly call the Pyramids after Philition, a shepherd, who at that time fed his flocks about the place." They thus called the Second, as well as the Great Pyramid, after him (iii. § 128); but, according to Professor Smyth, the Second Pyramid, though architecturally similar to the first, and almost equal in size, has nothing about it of the "superhuman" character of the Great Pyramid.

Messrs. Taylor and Smyth to have been hewn out and placed there as a measure of capacity for the world, so that the ancient Hebrew, Grecian, and Roman measures of capacity on the one hand, and our modern Anglo-Saxon on the other, are all derived, directly or indirectly, from the parent measurements of this granite vessel. "For," argues Mr. Taylor, "the porphyry coffer in the King's Chamber was intended to be a standard measure of capacity and weights for all nations; and all chief nations did originally receive their weights and measures from thence."

The works of these authors show, in numerous passages and extracts,[1] that, in their belief, the great object for which the whole pyramid was created, was the preservation of this coffer as a standard of measures, and the "whole pyramid arranged in subservience to it." The accounts of it published by Mr. Taylor, and in Mr. Smyth's first work, further aver that the coffer is, internally and externally, a rectangular figure of mathematical form, and of "exquisite geometric truth," "highly polished, and of a fine bell-metal consistency" (p. 99). "The chest or coffer in the Great Pyramid" (writes Mr. Taylor in 1859) "is so shaped as to be in every part rectangular from side to side, and from end to end, and the bottom is also cut at right angles to the sides and end, and made perfectly level." "The coffer," said Professor Smyth in 1864,

[1] The extracts within inverted commas, here, and in other parts, are from —(1.) Mr. John Taylor's work, entitled *The Great Pyramid—Why was it Built, and Who Built it?* London, 1859; and (2.) Professor Smyth's work, *Our Inheritance in the Great Pyramid*, Edinburgh, 1864; (3.) his later three-volume work, *Life and Work at the Great Pyramid*, Edinburgh, 1867; and (4.) *Recent Measures at the Great Pyramid*, in the Transactions of the Royal Society of Edinburgh for 1865–66.

"exhibits to us a standard measure of 4000 years ago, with the tenacity and hardness of its substance unimpaired, and the polish and evenness of its surface untouched by nature through all that length of time."

But later inquiries and observations upset entirely all these notions and strong averments in regard to the coffer. For—

(1.) *The Coffer, though an alleged actual standard of capacity-measure, has yet been found difficult or impossible to measure.*—In his first work, "Our Inheritance in the Great Pyramid," Professor Smyth had cited the measurements of it, made and published by twenty-five different observers, several of whom had gone about the matter with great mathematical accuracy.[1] Though imagined to be a great standard of measure, yet all these twenty-five, as Professor Smyth owned, varied from each other in their accounts of this imaginary standard in "every element of length, breadth, and depth, both inside and outside." Professor Smyth has latterly measured it himself, and this twenty-sixth measurement varies again from all the preceding twenty-five. Surely a measure of capacity should be measureable. Its mensurability indeed ought to be its most unquestionable quality; but this imagined standard has proved virtually unmeasurable—in so far at least that its twenty-six different and skilled measurers all differ from each other

[1] Professor Smyth has omitted to state—what, after all, it was perhaps unnecessary to state—that one set of these measurements, which he has tabulated and published, viz., that given by Dr. Whitman, was taken for him "by a British officer of engineers;" as, when Dr. Whitman visited Gizeh, he did not himself examine the interior of the Great Pyramid.—See Colonel Vyse's work, vol. ii. p. 286.

in respect to its dimensions. Still, says Professor Smyth, "this affair of the coffer's precise size is *the question of questions.*"

(2.) *Discordance between its actual and its theoretical measure.*—Professor Smyth holds that *theoretically* its capacity ought to be 71,250 "pyramidal" cubic inches, for that cubic size would make it the exact measure for a chaldron, or practically the vessel would then contain exactly four quarters of wheat, etc. Yet Professor Smyth himself found it some 60 cubic inches less than this; while also the measurements of Professor Greaves, one of the most accurate measurers of all, make it 250 cubic inches, and those of Dr. Whitman 14,000 *below* this professed standard. On the other hand, the measurements of Colonel Howard Vyse make it more than 100, those of Dr. Wilson more than 500, and those of the French academicians who accompanied the Napoleonic expedition to Egypt, about 6000 cubic inches *above* the theoretical size which Professor Smyth has latterly fixed on.

(3.) *Its theoretical measure varied.*—The *actual* measure of the coffer has varied in the hands of all its twenty-six measurers. But even its *theoretical* measure is varied also; for the size which the old coffer really *ought* to have as "a grand capacity standard," is, strangely enough, not a determined quantity. In his last work (1867), Professor Smyth declares, as just stated, its proper theoretical cubic capacity to be 71,250 pyramidal cubic inches. But in his first work (1864), he declared something different, for "we *elect,*" says he, "to take 70,970·2 English cubic inches (or 70,900 pyramidal cubic inches) as the true, because the theoretically *proved* contents of the porphyry coffer, and therefore accept these numbers as giving the cubic size of the grand *standard* measure of capacity in the Great Pyramid. Again, however, Mr. Taylor, who, previously to Professor Smyth, was the great advocate of the coffer being a marvellous standard of capacity measure for all nations, ancient and modern, declares its measure to be neither of the above quantities, but 71,328 cubic inches, or a cube of the ancient cubit of Karnak.[1] A vessel cannot be a measure of capacity whose own standard theoretical size is thus declared to vary somewhat every few years by those very men who maintain that it is a standard. But whether its capacity is 71,250, or 70,970, or 71,328, it is quite equally held up by Messrs Taylor and Smyth that the Sacred Laver of the Israelites, and the Molten Sea of the Scriptures, also conform and correspond to its (yet undetermined) standard "with *all* conceivable practical exactness;" though the standard of capacity to which they thus conform and correspond is itself a size or standard which has not been yet fixed with any exactness. Professor Smyth, in speaking of the calculations and theoretical dimensions of this coffer —as published by Mr. Jopling, a believer in its wonderful standard character—critically and correctly observes, "Some very astonishing results were brought out in the play of arithmetical numerations."

(4.) *The dilapidation of the Coffer.*—Thirty years ago this stone coffer was pointed out, and indeed delineated by Mr. Perring, as "*not* particularly well polished," and "chipped and broken at the

[1] "Its contents," says Mr. Taylor (p. 299), "are equal in cubic inches to the cube of 41,472 inches—the cubit of Karnak—viz., to 71,328 cubic inches." Elsewhere (p. 304) he states—"The Pyramid coffer contains 256 gallons of wheat;"—"It also contains 256 gallons of water, etc."

edges." Professor Smyth, in his late travels to Egypt, states that he found every possible line and edge of it chipped away with large chips along the top, both inside and outside, "chip upon chip, woefully spoiling the original figure; along all the corners of the upright sides too, and even along every corner of the bottom, while the upper south-eastern corner of the whole vessel is positively broken away to a depth and breadth of nearly a third of the whole." Yet this broken and damaged stone vessel is professed to be the *permanent* and perfect miraculous standard of capacity-measure for the world for "present and still future times;" and, according to Mr. Taylor—that it might serve this purpose, "is formed of one block of the hardest kind of material, such as porphyry or granite, *in order* that it might *not* fall into decay;" for "in this porphyry coffer we have" (writes Professor Smyth in 1864) "the very closing end and aim of the whole pyramid."

(5.) *Alleged mathematical form of the Coffer erroneous.*—But in regard to the coffer as an exquisite and marvellous standard of capacity to be revealed in these latter times, worse facts than these are divulged by the tables, etc., of measurements which Professor Smyth has recently published of this stone vessel or chest. His published measurements show that it is not at all a vessel, as was averred a few years ago, of pure mathematical form; for, externally, it is in length an inch greater on one side than another; in breadth half-an-inch broader at one point than at some other point; its bottom at one part is nearly a whole inch thicker than it is at some other parts; and in thickness its sides vary in some points about a quarter of an inch near the top. "But," Professor Smyth adds, "if calipered lower down, it is extremely probable that a *notably* different thickness would have been found there ;"—though it does not appear why they were not thus calipered.[1] Further, externally, "all the sides" (says Professor Smyth) "were slightly hollow, excepting the east side ;" and the "two external ends" also show some "concavity" in form. "The outside," (he avows) "of the vessel was found to be by no means so perfectly accurate as many would have expected, for the length was greater on one side than the other, and *different* also according to the height at which the measure was made." "The workmanship" (he elsewhere describes) "of the *inside* is in advance of the outside, but yet *not* perfect." For internally there is a convergence at the bottom towards the centre; both in length and in breadth the interior differs about half-an-inch at one point from another point; the "extreme points" (also) "of the corners of the bottom not being perfectly worked out to the intersection of the general planes of the entire sides ;" and thus its cavity seems really of a form utterly unmeasurable in a correct way by mere linear measurement—the only measure yet attempted. If it were an object of the slightest moment, perhaps liquid measurements would be more successful in ascertaining at least as much of the mensuration of the lower part of the coffer as still remains.

(6.) *Coffer cut with ledges and catch-holes for a lid, like other sarcophagi.*—More damaging details still remain in relation to the coffer as "a grand standard measure of capacity," and prove that

[1] At a later meeting of the Royal Society, on 20th April, Professor Smyth explained that, among the numerous instruments he carried out, he was not provided with calipers fit for this measurement.

its object or function was very different. In his first work Professor Smyth describes the coffer as showing no "symptoms" whatever of grooves, or catchpins or other fastenings for a lid. "More modern accounts," he re-observes, "have been further precise in describing the smooth and geometrical finish of the upper part of the coffer's sides, *without any* of those grooves, dovetails, or steady-pin-holes which have been found elsewhere in true polished sarcophagi, where the firm fastening of the lid is one of the most essential features of the whole business." Mr. Perring, however, delineated the catchpin-holes for a lid in the coffer thirty years ago.[1] On his late visit to it Professor Smyth found its western side lowered down in its whole extent to nearly an inch and three-quarters (or more exactly, 1·72 inch), and ledges cut out around the interior of the other sides at the same height. Should we measure on this western side from this actual ledge brim, or from the imaginary higher brim? If reckoned as the true brim, "this ledge" (according to Professor Smyth) would "take away near 4000 inches from the cubic capacity of the vessel." Besides, he found three holes cut on the top of the coffer's lowered western side, as in all the other Egyptian sarcophagi, where these holes are used along with the ledge and grooves to admit, and form a simple mechanism to lock the lids of such stone chests.[2]

[1] See plate iii. Fig. 1, in his great folio work on the *Pyramids of Gizeh from Actual Survey and Admeasurement*, Lond. 1839. "The sarcophagus is," he remarks, "of granite, not particularly well polished; at present it is chipped and broken at the edges. There are not any remains of the lid, *which was however*, fitted on in the same manner as those of the other pyramids."

[2] "The western side," observes Professor Smyth, "of the coffer is, through almost its entire length, rather lower than the other three, and these have *grooves* inside, or the remains of grooves once cut into them, about an inch or two below their summits, and on a level with the western edge; *in fact*, to

In other words, it presents the usual ledge and apparatus pertaining to Egyptian stone sarcophagi, and served as such.

(7.) *Sepulchral contents of Coffer when first discovered.*—When, about a thousand years ago, the Caliph Al Mamoon tunnelled into the interior of the pyramid, he detected by the accidental falling, it is said, of a granite portcullis, the passage to the King's Chamber, shut up from the building of the pyramid to that time. "Then" (to quote the words of Professor Smyth) "the treasures of the pyramid, sealed up almost from the days of Noah, and undesecrated by mortal eye for 3000 years, lay full in their grasp before them." On this occasion, to quotethe words of Ibn Abd Al Hakm or Hokm—a contemporary Arabian writer, and a historian of high authority,[1] who was born, lived, and died in Egypt—they found in the pyramid, "towards the top, a chamber [now the so-called King's Chamber] with an hollow stone [or coffer] in which there was a statue [of stone] like a man, and within it a man upon whom was a breastplate of gold set with jewels; upon this breastplate was a sword of inestimable price, and at his head a carbuncle of the bigness of an egg, shining like the light of the day; and upon him were characters writ with a pen,[2] which no man understood"[3]—a

admit a sliding sarcophagus cover or lid; and there were the remains of three fixing pin-holes on the western side, for fastening such cover into its place." (Vol. i. p. 85.)

[1] For age, etc., of Al Hakm, see Dr. Rieu in APPENDIX No. III.; and Jomard on length of the Sarcophagus, No. IV.

[2] In the original Arabic, the expression is "birdlike (or hieroglyphic) characters writ with a reed."

[3] See Greaves' *Works*, vol. i. p. 61 and p. 115. In Colonel Vyse's works are adduced other Arabian authors who allude to this discovery of a body with

description stating, down to the so-called "statue," mummy-case, or cartonage, and the hieroglyphics upon the cere-cloth, the arrangements now well known to belong to the higher class of Egyptian mummies.

In short (to quote the words of Professor Smyth), "that wonder within a wonder of the Great Pyramid—viz., the porphyry coffer," —that "chief mystery and boon to the human race which the Great Pyramid was built to enshrine,"—"this vessel of exquisite meaning," and of "far-reaching characteristics,"—mathematically formed under alleged Divine inspiration as a measure of capacity (and, according to M. Jomard, probably of length also) for all men and all nations, for all time,—and particularly for these latter profane times,—is, in simple truth, nothing more and nothing less than—an old and somewhat misshapen stone coffin.

Standard of Linear Measure in the Great Pyramid.

The standard in the Great Pyramid, according to Messrs. Taylor and Smyth, for *linear* measurements, is the length of the base line or lines of the pyramid. This, Professor Smyth states, is "*the function proper of the pyramid's base.* It is professed also that in this base line there has been found a new mythical inch—one-

golden armour, etc., etc., in the sarcophagus of the King's Chamber; as Alkaisi, who testifies that "he himself saw the case (the cartonage or mummy-case) from which the body had been taken, and that it stood at the door of the King's Palace at Cairo, in the year 511" A.H. (See *The Pyramids of Gizeh*, vol. ii. p. 334. See also to the same effect *Abou Szalt*, p. 357; and Ben Abd Al Rahman, as cited in the *Description de l'Egypte*, vol. ii. p. 191. "It may be remarked," observes Dr. Sprenger in Colonel Vyse's work, "that the Arabian authors have given the same accounts of the pyramids, with little or no variation, for above a thousand years." (Vol. ii. p. 328.) See further Appendix, p. 270.

thousandth of an inch longer than the British standard inch ; and in the last sections of his late work Professor Smyth has earnestly attempted to show that the status of the kingdoms of Europe in the general and moral world may be measured in accordance with their present deviation from or conformity to this suppositious pyramidal standard in their modes of national measurement.[1] "For the linear measure" (says Professor Smyth) "of the base line of this colossal monument, viewed in the light of the philosophical connection between time and space, has yielded a standard measure of length which is more admirably and learnedly earth-commensurable than anything which has ever yet entered into the mind of man to conceive, even up to the last discovery in modern metrological science, whether in England, France, or Germany."

The engineers and mathematicians of different countries have repeatedly measured arcs of meridians to find the form and dimensions of the earth, and the French made the metre (their standard of length), $\frac{1}{10,000,000}$ of the quadrant of the meridian. Professor Smyth holds that the basis line of the pyramid has been laid down by Divine authority as such a guiding standard measure.

What, then, is the exact length of one of its basis lines? The sides of the pyramid have been measured by many different measurers. Linear standards have, says Professor Smyth, "been already looked for by many and many an author on the sides of the base of the Great Pyramid, even before they knew that the terminal points of those magnificent base lines had been carefully marked in the solid rock of the hill by the socket-holes of the

[1] See Appendix, No. VII.

builders." But—as in the case of the cubic capacity of the coffer—these measurers sadly disagree with each other in their measurements, which, in fact, vary from some 7500 or 8000 inches to 9000 and upwards. Thus, for example, Strabo makes it under 600 Grecian feet, or under 7500 English inches ; Dr. Shawe makes it 8040 inches ; Shelton makes it 8184 inches ; Greaves, 8316 ; Davison, 8952 : Caviglia, 9072 ; the French academicians, 9163 ; Dr. Perry, 9360, etc., etc.

At the time at which Professor Smyth was living at the Pyramids, Mr. Inglis of Glasgow visited it, and, for correct measurement, laid bare for the first time the four corner sockets. Mr. Inglis's measurements not only differed from all the other measurements of "one side" base lines made before him, but he makes the four sides differ from each other ; one of them—namely, the north side—being longer than the other three. Strangely, Professor Smyth, though in Egypt for the purpose of measuring the different parts of the pyramid—and holding that its base line ought to be our grand standard of measure, and further holding that the base line could only be accurately ascertained by measuring from socket to socket—never attempted that linear measurement himself after the sockets were cleared. These four corner sockets were never exposed before in historic times ; and it may be very long before an opportunity of seeing and using them again shall ever be afforded to any other measurers.

Before the corner sockets were exposed, Professor Smyth attempted to measure the bases, and made each side of the present masonry courses "between 8900 and 9000 inches in length," or (to use his own word) "*about*" 8950 inches for the mean length of one of the four sides of the base ; exclusive of the ancient casing and backing stones—which last Colonel Howard Vyse found and measured to be precisely 108 inches on each side, or 216 on both sides. These 216 inches, added to Professor Smyth's measure of "about" 8950 inches, make one side 9166 inches. But Professor Smyth has "elected" (to use his own expression) not to take the mathematically exact measure of the casing stones as given by Colonel Vyse and Mr. Perring, who alone ever saw them and measured them (for they were destroyed shortly after their discovery in 1837), but to take them, without any adequate reason, and contrary to their mathematical measurement, as equal only to 202 inches, and hence "accept 9152 inches as the original length of one side of the base of the finished pyramid." He deems, however, this "determination" not to be so much depended upon as the measurements made from socket to socket.

The mean of the only four series of such socket or casing stone measures as have been recorded hitherto by the French Academicians (9163), Vyse (9168), Mahmoud Bey (9162), and Inglis (9110), amounts to nearly 9150. The first three of these observers were only able to measure the north side of the pyramid. Mr. Inglis measured all the four sides, and found them respectively 9120, 9114, 9102, and 9102, making a difference of 18 inches between the shortest and longest. Professor Smyth thinks the measures of Mr. Inglis as on the whole probably too *small*, and he takes two of them, 9114 and 9102—(but, strangely, not the largest, 9120)—as data, and strikes a new number out of these two, and out of the three previous measures of the French Academicians, Vyse, and Mahmoud Bey; from these five quantities making a calculation of "means," and electing 9142 as the proper measure of the basis line of the pyramid—(which exact measure certainly none of its many measurers ever yet found it to be); and upon this *foundation*, "derived" (to use his own words) "from the best modern measures yet made," he proceeds to reason, "as the happy, useful, and perfect representation of 9142," and the great standard for linear measure revealed to man in the Great Pyramid. Surely it is a remarkably strange *standard* of linear measure that can only be thus elicited and developed—not by direct measurement but by indirect logic; and regarding the exact and precise length of which there is as yet no kind of reliable and accurate certainty.

Lately, Sir Henry James, the distinguished head of the Ordnance Survey Department, has shown that the length of one of the sides of the pyramid base, with the casing stones added, as measured by Colonel H. Vyse—viz. 9168 inches—is precisely 360 derahs, or land cubits of Egypt; the derah being an ancient land measure still in use, of the length of nearly 25½ British inches, or, more correctly, of 25·488 inches; and he has pointed out that in the construction of the body of the Great Pyramid, the architect built 10 feet or 10 cubits of horizontal length for every 9 feet or 9 cubits of vertical height; while in the construction of the inclined passages the proportion was adhered to of 9 on the incline to 4 in vertical height, rules which would altogether simplify the building of such a structure.[1] The Egyptian derah of 25·48 inches is practically one-fourth more in length than the old cubit of the city of Memphis. Long ago Sir Isaac Newton showed, from Professor Greaves' measurements

[1] Our great Scottish architect, Mr. Bryce, believes that, with these data given, any well-informed master-mason or clerk of works could have drawn or planned and superintended the building.

of the chambers, galleries, etc., that the Memphis cubit (or cubit of "ancient Egypt generally") of 1·719 English feet,[1] or 20·628 English inches, was apparently the *working* cubit of the masons in constructing the Great Pyramid[2]—an opinion so far admitted more lately by both Messrs Taylor and Smyth; "the length" (says Professor Smyth) "of the cubit employed by the masons engaged in the Great Pyramid building, or that of the ancient city of Memphis," being, he thinks, on an average taken from various parts in the interior of the building, 20·73 British inches.[3] According to Mr. Inglis' late measurement of the four bases of the pyramid, after its four corner sockets were exposed, the length of each base line was possibly 442 Memphis cubits, or 9117 English inches; or, if the greater length of the French Academicians, Colonel Vyse, and Mahmoud Bey, be held nearer the truth, 444 Memphis cubits, or 9158 British inches.

But Professor Smyth tries to show that (1.) if 9142 only be granted to him as the possible base line of the pyramid; and (2.) if 25 pyramidal inches be allowed to be the length of the "Sacred Cubit," as revealed to the Israelites (and as revealed in the pyra-

[1] See Newton's *Essay*, in Professor Smyth's work, vol. ii. 360; and Sir Henry James' masterly *Memorandum on the Length of the cubit of Memphis*, in APPENDIX, No. V.

[2] Sir Isaac Newton says—"In the precise determination of the cubit of Memphis, I should choose to pitch upon the length of the chamber in the middle of the pyramid." Greaves gives this length 34·38 = 20 cubits of 20·628 inches.

[3] Yet this, the Memphian cubit, "need not" (somewhat mysteriously adds Professor Smyth), "and actually is not, by any means the same as the cubit *typified* in the more concealed and *symbolised* metrological system of the Great Pyramid."

mid), then the base line might be found very near a multiple of this cubit by the days of the year,[1] or by 365·25 ; for these two numbers multiplied together amount to 9131 "pyramidal" inches, or 9140 British inches—the British inch being held, as already stated, to be 1000th less than the pyramidal inch. Was, however, the "Sacred Cubit"—upon whose alleged length of 25 "pyramidal" inches this idea is entirely built—really a measure of this length? In this matter—the most important and vital of all for his whole linear hypothesis—Professor Smyth seems to have fallen into errors which entirely upset all the calculations and inferences founded by him upon it.

Length of the Sacred Cubit.—Sir Isaac Newton, in his remarkable *Dissertation upon the Sacred Cubit of the Jews* (republished in full by Professor Smyth in the second volume of his *Life and Work at the Great Pyramid*), long ago came to the conclusion that it measured 25 unciæ of the Roman foot, and $\frac{6}{10}$ of an uncia, or 24·753 British inches; and in this way it was one-fifth longer than the cubit of Memphis—viz. 20·628 inches, as previously deduced by him

[1] Godfrey Higgins, in his work on *The Celtic Druids*, shows how, among the ancients, superstitions connected with numbers, as the days of the year or the figures 365, have played a prominent part. "Amongst the ancients" (says he) "there was no end of the superstitious and trifling play upon the nature and value of numbers. The first men of antiquity indulged themselves in these fooleries" (p. 244). Mr. Higgins points out that the old Welsh or British word for Stonehenge, namely Emrys, signifies, according to Davies, 365 ; as do the words Mithra, Neilos, etc. ; that certain collections of the old Druidic stones at Abury may be made to count 365 ; that "the famous Abraxas only meant the solar period of 365 days, or the sun," etc. "It was all judicial astrology. . . . . It comes" (adds Mr. Higgins) "from the Druids."

from Greaves' measurements of the King's Chamber and other parts of the interior of the Great Pyramid. Before drawing his final inference as to the Sacred Cubit being 24·75 inches, and as so many steps conducting to that inference, Sir Isaac shows that the Sacred Cubit was some measurement intermediate between a long and moderate human step or pace, between the third of the length of the body of a tall and short man, etc. etc. Professor Smyth has collected several of the estimations thus adduced by Newton as "methods of approach" to circumscribe the length of the Sacred Cubit, and omitted others. Adding to eight of these alleged data, what he mistakingly avers to be Sir Isaac's deduction of the actual length of the Sacred Cubit in British inches—(namely, 24·82 instead of 24·753)—as a ninth quantity, he enters the whole nine in a table as follows :—

Professor Smyth's Table of Newton's data of Inquiry regarding the Sacred Cubit.[1]

"First	between	23·28 and	27·94	British inches.
Second	„	23·3	27·9	„
Third	„	24·80	25·02	„
Fourth	„	24·91	25·68[2]	„
And Fifth, somewhere near		24·82."		

"The mean of all which numbers" (Professor Smyth remarks)

[1] See this table in Professor Smyth's *Life and Work at the Great Pyramid*, vol. ii. p. 458. The table professes to give some of Sir Isaac Newton's data regarding the Sacred Cubit by changing the measurements which Sir Isaac uses of the Roman foot and inch into English inches. But all the figures and measurements are transferred into English inches by a different rule from that

[2] The fourth line in the table presents a most fatal and unfortunate error in a special calculation to which the very highest importance is professed to

"amounts to 25·07 British inches. The Sacred Cubit, then, of the Hebrews" (he adds) "in the time of Moses—*according to Sir Isaac Newton*—was equal to 25·07 British inches, with a probable error of ± ·1."

But—"*according* to Sir Isaac Newton"—the Sacred Cubit of the Jews was *not* 25·07, as Professor Smyth makes him state in this table, but 24·75 British inches, as Sir Isaac himself more than once deliberately infers in his Dissertation.[1] Besides, in such

which Sir Isaac himself lays down—viz., that the English foot is 0·967 of the Roman foot; and, consequently, *in every one of the instances given* in Mr. Smyth's table, the lengths in English inches of these data of Sir Isaac Newton are assuredly *not* their lengths in English inches as understood and laid down by Newton himself.

[1] Thus, after deducing the length of the cubit of Memphis from the length of the King's Chamber, Sir Isaac Newton observes:—"From hence I would infer that the Sacred Cubit of Moses was equal to 25 unciæ of the Roman foot and $\frac{6}{10}$ of an *uncia*." (See his *Dissertation on the Sacred Cubit*, as republished in Professor Smyth's *Life and Work at the Great Pyramid*, vol. ii. p. 362.) Again, at p. 363, Sir Isaac speaks of "the cubit which we have concluded to have been in the time of Moses $25\frac{60}{100}$ inches" of the Roman foot; and at p. 365, in closing his Dissertation, he remarks—"The Roman cubit therefore consists of 18 unciæ, and the Sacred Cubit of $25\frac{3}{5}$ unciæ, of the Roman foot." In other words, according to Sir Isaac Newton, the Sacred

---

be attached. This fourth line gives the measurement of the Sacred Cubit as quoted by Newton from Mersennus, who laid down its length as 25·68 inches of Roman measurement. Professor Smyth changes this Roman measurement into 24·91 English inches, and then erroneously enters these same identical Roman and English measurements of Mersennus—viz., 24·91 and 25·68—not as *one* identical quantity, which they are—but as *two* different and contrasting quantities; and further, he tabulates this strange mistake as one of the "methods of approach" for gaining a correct idea of the Sacred Cubit. Never, perhaps, has so unhappy an error been made in a work of an arithmetical and mathematical character.

inquiries, is it not altogether illogical to attempt to draw mathematical deductions by these calculations of "means," and especially by using the ninth quantity in the table—viz. Sir Isaac's own avowed and deliberate deduction regarding the actual length of the Sacred Cubit—as one of the nine quantities from which that length was to be again deduced by the very equivocal process of "means?" Errors, however, of a far more serious kind exist. The "mean" of the nine quantities in Professor Smyth's table is, he infers, 25·07 inches; and hence he avows that this, or near this figure, is the length of the Sacred Cubit. But the real mean of the nine quantities which Professor Smyth has collected is not 25·07 but **25·29**—a number in such a testing question as this of a very different value. For the days of the year (365·25) when multiplied by this, the true mean of these nine quantities, would make the base line of the pyramid 9237 inches instead of Professor Smyth's theoretical number of 9142 inches; a difference altogether overturning all his inferences and calculations thereanent. And again, if we take Sir

Cubit of 25·60 inches of the Roman foot is equal to 24·75 British inches; for, as he calculated, the Roman foot "was equal to $\frac{967}{1000}$ of the English foot." (See p. 342.) This is the measurement of the Roman foot laid down by Sir Isaac Newton in his Dissertation, and the only standard of it mentioned in Professor Smyth's *Life and Work at the Great Pyramid;* yet in that work Professor Smyth calculates Sir Isaac's Sacred Cubit to be 24·82 instead of 24·75 British inches. In doing so, he has calculated the English foot as equal to ·970 of the Roman foot; but was he entitled to do so when using Sir Isaac's own data, and when employing Sir Isaac's own calculated conclusion as to the length of the Sacred Cubit? In the published *Proceedings* of the Royal Society, in consequence of following the calculation by Professor Smyth of Sir Isaac Newton's conclusion from Sir Isaac's own data as to the length of the Sacred Cubit, it was erroneously spoken of as 24·82, instead of 24·75 British inches.

Isaac Newton's own conclusion of 24·75, and multiply it by the days of the year, the pretended length of the pyramid base comes out as low as 9039.

Alleged "really glorious Consummation" in Geodesy.

The incidentally but totally erroneous summation which Professor Smyth thus makes of the nine equivocal quantities in his table, as amounting to 25·07, he declares (to use his own strong words) as a "*really glorious consummation* for the geodesical science of the present day to have brought to light;" for he avers this length of 25·07—(which he forthwith elects to alter and change, without any given reason whatever, to 25·025 British inches)—being, he observes, "practically the sacred Hebrew cubit, is *exactly* one ten-millionth (1--10,000,000th) of the earth's semi-axis of rotation; and *that* is the very best mode of reference to the earth-ball as a whole, for a linear standard through all time, that the highest science of the existing age of the world has yet struck out or can imagine. In a word, the Sacred Cubit, *thus* realised, forms an instance of the most advanced and perfected human science supporting the truest, purest, and most ancient religion; while a linear standard which the chosen people in the earlier ages of the world were merely told by maxim to look on as *sacred*, compared with other cubits of other lengths, is proved by the progress of human learning in the latter ages of time, to have had, and still to have, a philosophical merit about it which no men or nations at the time it was first produced, or within several thousand years thereof, could have possibly thought of for themselves." Besides, adds he elsewhere, "an *extraordinarily*[1] convenient length too, for man to handle and use in the common affairs of life is the one ten-millionth of the earth's semi-axis of rotation when it comes to be realised, for it is extremely close to the ordinary human arm, or to the ordinary human pace in walking, with a purpose to measure."

Of course all these inferences and averments regarding the Sacred Cubit being an exact segment of the polar axis disappear, when we find Sir Isaac Newton's length of the Sacred Cubit is not, as Professor Smyth elects it to be, 25·025 British inches; nor 25·07, as he incorrectly calculated it to be from the mean of the nine quantities selected and arranged in his table; nor 25·29, as is the actual mean of these nine quantities in his table; but, "*according* to Sir Isaac Newton's" own reiterated statement and conclusion, 24·753. (See footnote, p. 245.) A Sacred Cubit, according to Sir Isaac Newton's admeasurements of it, of 24·75 inches, would not, by thousands of cubits, be one ten-millionth of the measure of the semi-polar axis of the earth; provided the polar axis be, as Professor Smyth elects it to be, 500,500,000 British inches.[2]

[1] This word "extraordinarily," was, by a clerical or printer's error, spelled "extraordinary" in the *Proceedings* of the Royal Society; and a friend who looked over the printed proof, and suggested two or three corrections, placed the word (sic) on the margin after it, from whence it slipped into the text:—accidents to be much regretted, as, from Professor Smyth's remarks to the Society on the 20th April, they had evidently given him much, but most unintentional offence.

[2] At the close of a subsequent meeting of the Royal Society, on the 20th April 1868, Professor Smyth gave away a printed Appendix to his three-volume work, in which he has acknowledged the erroneous character—as pointed out in this communication—of his all-important table, p. 22, on the length of the Sacred Cubit, by withdrawing it, and offering one of a new construction and character, but without being able to make the length of the cubit come nearer to his theory. See further, Appendix, No. VI.

Axis of the Earth as a Standard of Measure.

The standards of measure in France and some other countries are, as is well known, referred to divisions of arcs of the meridian, measured off upon different points of the surface of the earth. These measures of arcs of the meridian, as measurements of a known and selected portion of the surface of the spheroidal globe of the earth, have, more or less, fixed mathematical relations with the axis of the earth; as the circumference of a sphere has an exact mathematical ratio to its diameter. The difference in length of arcs of the meridian at different parts of the earth's surface, in consequence of the spheroidal form of the globe of the earth, has led to the idea that the polar diameter or axis of the earth would form a more perfect and more universal standard than measurements of the surface of the earth. In the last century, Cassini[1] and Callet[2] proposed, on these grounds, that the polar axis of the earth should be taken as the standard of measure. Without having noticed these propositions of Cassini and Callet,[3] Professor Smyth adopts the same idea, and avers that 4000 years ago it had been adopted and used also by the builders of the Great Pyramid, who laid out and measured off the basis of the pyramid as a multiple by the days of the year of the Sacred Cubit, and hence of the Pyramidal Cubit while the Sacred or Pyramidal Cubit were both the results of super-

[1] *Traite de la Grandeur et de la Figure de la Terre.* Amsterdam edition (1723), p. 195.

[2] *Tables Portatives de Logarithmes.* Paris, 1795, p. 100.

[3] The same idea of using the earth's axis as a standard of length has been suggested also by Professor Hennessy of Dublin, and by Sir John Herschel See *Athenæum* for April 1860, pp. 581 and 617.

human or divine knowledge, and were both, or each, one ten-millionth of the semi-polar axis of the earth. We have already seen, however, that the Sacred Cubit, "*according* to Sir Isaac Newton," is not a multiple by the days of the year of the base line of the Great Pyramid; and is not one twenty-millionth of the polar axis of the earth, when that polar axis is laid down as measuring, according to the numbers elected by Professor Smyth, 500,500,000 British inches.

But is there any valid reason whatever for fixing and determining, as an ascertained mathematical fact, the polar axis of the earth to be this very precise and exact measure, with its formidable tail of cyphers? None, except the supposed requirements or necessities of Professor Smyth's pyramid metrological theory. The latest and most exact measurements are acknowledged to be those of Captain Clarke, who, on the doctrine of the earth being a spheroid of revolution computes the polar axis to be 500,522,904 British inches, calculating it from the results of all the known arcs of meridian measures. If we grant that the Sacred Cubit could be allowed to be exactly 25·025 inches, which Sir Isaac Newton found it not to be; and if we grant that the polar axis is exactly 500,500,000 British inches, which Captain Clarke did not find it to be; then, certainly, as shown by Professor Smyth, there would be 20,000,000 of these supposititious pyramidal cubits, or 500,000,000 of the supposititious pyramidal inches in this supposititious polar axis of the earth. "In so far, then" (writes Professor Smyth), "we have in the 5, with the many 0's that follow, a pyramidally commensurable and symbolically appropriate unit for the earth's axis of rotation." But such adjustments have been made with as great apparent exactitude when entirely different earth-axes and quantities were taken. Thus Mr. John Taylor shows the inches, cubits, and axes to answer precisely, although he took as his standard a totally different diameter of the earth from Professor Smyth. The diameter of the earth at 30° of latitude—the geographical position of the Great Pyramid—is, he avers, some seventeen miles, or more exactly 17·652 miles longer than at the poles.[1] But Mr. Taylor fixed upon this diameter of the earth at latitude 30°—and not, like Professor Smyth, upon its polar diameter—as the standard for the metrological linear measures of the Great Pyramid; and yet, though the standard was so different, he found, like Mr. Smyth, 500,000,000 of inches also in his axis, and 20,000,000 of cubits also.[2] The resulting figures appear to fit equally as well for the one as for the other. Perhaps they answer best on Mr Taylor's scheme. For Mr. Taylor maintained that the diameter of the earth before the Flood, at this selected point of 30°, was less by nearly 37 miles than what it was subsequently to the flood,[3] and is now; a point by which he accounts

[1] The diameter of the earth in latitude 30° is really about 20 miles longer than the polar axis. But Mr. Taylor obviously did not know the nature of the spheroidal arcs of the meridian, and so falls into the most inconsistent assertions respecting the length of this particular diameter. Thus, in pp. 75 and 87, he asserts the diameter in latitude 30° to be 500,000,000 inches [that is =7891·414 miles], which is 7·756 miles *less* than the polar axis—*the least* diameter of all; whereas, in p. 95, he states this diameter in lat. 30° to be 17·652 miles *greater* than the polar axis.

[2] "The diameter of the earth, according to the measures taken at the Pyramids, is 41,666,667 English feet, or 500,000,000 inches." (See *The Great Pyramid*, p. 75.) "Dividing this number by 20,000,000 we obtain the measure of 25 (English) inches for the Sacred Cubit" (p. 67).

[3] "When" (says Mr. Taylor, p. 91) "the *new* Earth was measured in

for otherwise unaccountable circumstances in the metrological doctrines which have been attempted to be connected with the Great Pyramid. For while Mr. Taylor believes the Sacred Cubit to be 24·88, or possibly 24·90 British inches, he holds the new Pyramidal cubit to be 25 inches in full; and the Sacred and Pyramidal cubits to be different therefore from each other, though both inspired. In explanation of this startling difference in two measures supposed to be equally of sacred[1] origin, Mr. Taylor observes—"The smaller

Egypt after the Deluge, it was found that it exceeded the diameter of the *old* Earth by the difference between 497,664,000, inches and 500,000,000 inches; that is, by 2,336,000 inches, equal to 36·868 miles."

[1] *Alleged Sacred Character of the Scottish Yard or Ell Measure.*—Professor Smyth tries to show (iii. 597), that if Britain stands too low in his metrological testing of the European kingdoms and races, its "low entry is due to accepting the yard for the country's popular measure of length." But long ago the "divine" origin of the Scottish ell—as in recent times the divine origin of the so-called pyramidal cubit and inch—was pleaded rather strenuously. For when, in the 13th century, Edward I. of England laid before Pope Boniface his reasons for attaching the kingdom of Scotland to the Crown of England, he maintained, among other arguments, the justice and legality of this appropriation on the ground that his predecessor King Athelstane, after subduing a rebellion in Scotland under the auspices of St. John of Beverley, prayed that through the intervention of that saint, it "might be granted to him to receive a visible and tangible token by which all future ages might be assured that the Scots were rightfully subject to the King of England. His prayer was granted in this way: Standing in front of one of the rocks at Dunbar, he made a cut at it with his sword, and left a score which proved to be the *precise* length of an ell, and was adopted as the regulation test of that measure of length." This legend of the "miraculously created ellwand standard" was afterwards duly attested by a weekly service in the Church of St John of Beverley. (See Burton's *History of Scotland*, ii. 319.) In the official account of the miracle, as cited by Rymer, it is declared that during its performance the rock cut like butter or soft mud under the stroke of Athelstane's sword. "Extrahens gladium de vaginâ percussit in cilicem, quæ adeo

(24·88 is the Sacred Cubit which measured the diameter of the Earth *before* the Flood; the one by which Noah measured the Ark, as tradition says; and the one in accordance with which all the interior works of the Great Pyramid were constructed.[1] The larger (25) is the Sacred Cubit of the *present* Earth, according to the standard of the Great Pyramid when it was completed."

Surely such marked diversities and contradictions, and such strange hypothetical adjustments and re-adjustments of the data and calculations, entirely upset the groundless and extraordinary theory of the base of the pyramid being a standard of linear measurement; or a segment of any particular axis of the earth; or a standard for emitting a system of new inches and new cubits;—seeing, on the one hand, more particularly, that the basis line of the pyramid is still itself an unknown and undetermined linear quantity, as is also the polar axis of the earth of which it is declared and averred to be an ascertained, determined, and measured segment.

M. Paucton, in 1780, wrote a work in which he laid down the base side of the pyramid as 8754 inches; maintained, like Mr. Taylor and Mr. Smyth, that this length was a standard of linear measures; found it to be the measure of a portion of a degree of the meridian, such degree being itself the 360th part of a circle;—and apparently the calculations and figures answered as well as when the measurement was declared to be 9142 inches, and the line not a segment of

penetrabilis, Dei virtute agente, fuit gladio, quasi câdem horâ lapis butirum esset, vel mollis glarea; . . . et usque ad presentem diem, evidens signum patet, quod Scoti, ab Anglis devicti ac subjugata; monumento tali evidenter cunctis adeuntibus demonstrante." (Foedera, tom. i. pars ii. 771.)

[1] Elsewhere (p. 45) Mr. Taylor corroborates Sir Isaac Newton's opinion that the *working* cubit by which the Pyramid was built was the cubit of Memphis.

an arc of the circumference of the earth, but a segment of the polar axis of the earth; for De l'Isle lauds Paucton's meridian degree theory as one of the wondrous efforts of human genius, or (to use his own words) "as one of the chief works of the human mind!" Yet the errors into which Paucton was seduced in miscalculating the base line of the Pyramid as 8754 inches, and the other ways he was misled, are enough—suggests Professor Smyth—"to make poor Paucton turn in his grave."

Significance of Cyphers and Fives.

M. Paucton, Mr. Taylor, and those who have adopted and followed their pyramid metrological ideas, seem to imagine that if, by multiplying one of their measures or objects, they can run the calculation out into a long tail of terminal 0's, then something very exact and marvellous is proved. "When" (upholds Mr. Taylor), "we find in so complicated a series of figures as that which the measures of the Great Pyramid and of the Earth require for their expression, *round numbers* present themselves, or such as leave no remainder, we may be sure we have arrived at *primitive* measures." But many small and unimportant objects, when thus multiplied sufficiently, give equally startling strings of 0's. Thus, if the polar axis of the earth be held as 500,000,000 inches, and Sir Isaac Newton's "Sacred Cubit" be held, as Professor Smyth calculated it to be, viz. 24·82 British inches—then the long diameter of the brim of the lecturer's hat, measuring 12·4 inches, is 1-40,000,000th of the earth's polar axis; a page of the print of the Society's Transactions is 1-60,000,000th of the same; a print page of Professor Smyth's book, 6·2 inches in length, is 1-80,000,000th of this "great standard;" etc. etc. etc.

Professor Smyth seems further to think that the figure or number "five" plays also a most important symbolical and inner part in the configuration, structure, and enumeration of the Great Pyramid. "The pyramid" (says he) "embodies in a variety of ways the importance of five." It is itself "five-angled, and with its plane a five-sided solid, in which everything went by fives, or numbers of fives and powers of five." "With five, then, as a number, times of five, and powers of five, the Great Pyramid contains a mighty system of consistently subdividing large quantities to suit human happiness." To express this, Mr. Smyth suggests the new noun "fiveness." But it applies to many other matters as strongly, or more strongly than to the Great Pyramid. For instance, the range of rooms belonging to the Royal Society is "five" in number; the hall in which it meets has five windows; the roof of that hall is divided into five transverse ornamental sections; and each of these five transverse sections is subdivided into five longitudinal ones; the books at each end of the hall are arranged in ten rows and six sections—making sixty, a multiple of five; the official chairs in the hall are ten in number, or twice five; the number of benches on one side for ordinary fellows is generally five; the office-bearers of the Society are twenty-five in number, or five times five; and so on. These arrangements were doubtless, in the first instance, made by the Royal Society without any special relation to "fiveness," or the "symbolisation" of five; and there is not the slightest ground for any belief that the apparent "fiveness" of anything in the Great Pyramid had a different origin.

GREAT MINUTENESS OF MODERN PRACTICAL STANDARDS OF GAUGES.

In all these "standards" of capacity and length alleged to exist about the Great Pyramid, not only are the theoretical and actual sizes of the supposed "standards" made to vary in different books —which it is impossible for an actual "standard" to do—but the evidences adduced in proof of the conformity of old or modern measures with them is notoriously defective in complete aptness and accuracy. Measures, to be true counterparts, must, in mathematics, be not simply "near," or "very near," which is all that is generally and vaguely claimed for the supposed pyramidal proofs, but they must be entirely and *exactly* alike, which the pyramidal proofs and so-called standards fail totally and altogether in being. Mathematical measurements of lines, sizes, angles, etc., imply exactitude, and not mere approximation; and without that exactitude they are not mathematical, and—far more—are they not "superhuman" and "inspired."

Besides, it must not be forgotten that our real *practical* standard measures are infinitely more refined and many thousand-fold more delicate than any indefinite and equivocal measures alleged to be found in the pyramid by even those who are most enthusiastic in the pyramidal metrological theory. At the London Exhibition in 1851, that celebrated mechanician and engineer, Mr. Whitworth, of Manchester, was the first to show the possibility of ascertaining by the sense of touch alone the one-millionth of an inch in a properly-adjusted standard of linear measure; and in his great establishment at Manchester they work and construct machinery and tools of all kinds with differences in linear measurements amounting to one ten-thousandth of an inch. The standards of the English inch, etc., made by him for the Government—and now used by all the engine and tool makers, etc., of the United Kingdom—lead to the construction of machinery, etc., to such minute divisions; and the adoption of these standards has already effected enormous saving to the country by bringing all measured metal machinery, instruments, and tools, wherever constructed and wherever afterwards applied and used, to the same identical series of mathematical and precise gauges.

The Sabbath, etc. Typified in the Pyramid.

The communication next discussed some others amongst the many and diversified matters which Professor Smyth fancifully averred to be typified and symbolised in the Great Pyramid.

One, for example, of the chambers in the Great Pyramid—the so-called Queen's Chamber—has a roof composed of two large blocks of stone leaning against each other, making a kind of slanting or double roof. This double roof, and the four walls of the chamber count six, and typify, according to Professor Smyth, the six days of the week, whilst the floor counts, as it were, a seventh side to the room, "nobler and more glorious than the rest," and typifying something, he conceives, of a "nobler and more glorious order"—namely, the Sabbath; it is surely difficult to fancy anything more strange than this strange idea.[1] In forming this theory liberties are also confessedly taken with the floor in order to make it duly

[1] The interior of any Scottish cottage, where the inside of the thatched or slated roof is left exposed by uncovered joists within, contains, on the same principle, six sides, and a seventh or the floor.

larger than the other six sides of the room, and to do so he theoretically lifts up the floor till it is placed higher than the very entrance to the chamber; for originally the floor and sides are otherwise too nearly alike in size to make a symbolic *seven*-sided room with one of the sides proportionally and properly larger than the other six sides. Yet Professor Smyth holds that, in the above typical way, he has "shown," or indeed "proved entirely," that the Sabbath had been heard of before Moses, and that thus he finds unexpected and confirmatory light of a fact which, he avers, is of "extraordinary importance, and possesses a ramifying influence through many departments of religious life and progress."

He believes, also, that the corner-stone—so frequently alluded to by the Psalmist and the Apostles as a symbol of the Messiah—is the head or corner-stone of the Great Pyramid, which, though long ago removed, may yet possibly, he thinks, be discovered in the Cave of Machpelah; though how, why, or wherefore it should have found its way to that distant and special locality is not in any way solved or suggested.

Great Pyramid alleged to be a Superhuman, and more or less an Inspired Metrological Erection.

Professor Smyth holds the Great Pyramid to be in its emblems, and intentions and work "superhuman;" as "not altogether of human origination; and in that case whereto" (he asks) "should we look for any human assistance to men but from Divine inspiration?" "Its metrology is," he conceives, "directed by a higher Power" than man; its erection "directed by the *fiat* of Infinite

Wisdom;" and the whole "built under the direction of chosen men divinely inspired from on high for this purpose."

If of this Divine origin, the work should be absolutely perfect; but, as owned by Professor Smyth, the structure is not entirely correct in its orientation, in its squareness, etc. etc.—all of them matters proving that it is human, and not superhuman. It was, Professor Smyth further alleges, intended to convey standards of measures to all times down to, and perhaps beyond, these latter days, "to herald in some of those accompaniments of the promised millennial peace and goodwill to all men." Hence, if thus miraculous in its forseen uses, it ought to have remained relatively perfect till now. But "what feature of the pyramid is there" (asks Professor Smyth) "which renders at once in its measurements in the present day its ancient proportions? None." If the pyramid were a miracle of this kind, then the Arabian Caliph Al Mamoon so far upset the supposititious miracle a thousand years ago—(of course he could not have done so provided the miracle had been truly Divine)—when he broke into the King's Chamber and unveiled its contents; inasmuch as the builders, according to Professor Smyth, intended to conceal its secrets for the benefit of these latter times, and for this purpose had left a mathematical sign of two somewhat diagonal lines or joints in the floor of the descending passage, by which secret sign or clue[1] some men or man in the far distant future,

[1] "The *clue* was not prepared for any immediate successors of the builders, but was intended, on the contrary, to endure to a most remote period. And it has so endured and served such a purpose even down to those our own days." (Professor Smyth's *Life and Work at the Great Pyramid*, vol. i. p. 157.) "The builders, or planners rather, of the Great Pyramid, did not leave their building without sure testimony to its chief secret; for there,

visiting the interior, should detect the entrance to the chambers; and which secret sign Professor Smyth himself was, as he believes, the first "man" to discover two years ago. The secret, however, thus averred to be placed there for the detection of the entrance to the interior chambers in these latter times, has been discovered some 1000 years at least too late for the evolution of the alleged miraculous arrangement. And in relation to the Great Pyramid, as to other matters, we may be sure that God does not teach by the medium of miracle anything that the unaided intellect of man can find out; and we must beware of erroneously and disparagingly attributing to Divine inspiration and aid, things that are imperfect and human.

The communication concluded by a series of remarks, in which it was pointed out that at the time at which the Great Pyramid was built, probably about 4000 years ago, mining, architecture, astronomy, etc., were so advanced in various parts of the East as to present no obstacle in the way of the erection of such magnificent mausoleums, as the colossal Great Pyramid and its other congener pyramids undoubtedly are.

before the eyes of all men for ages, had existed these *two diagonal joints* in the passage floor, pointing directly and constantly to what was concealed in the roof just opposite them, and no one ever thought of it. Practically, then, we may say with full certainty that these two floor marks were left there to guide *men* who, it was expected, would come subsequently, earnestly desiring, on rightly-informed principles, to look for the entrance to the upper parts of the Pyramid." (Vol. i. p. 156-7.) At p. 270 Professor Smyth again alludes to this supposed mark, made up by two diagonal joints in the passage floor, as evading the notice of all visitors, except "those very few, or perhaps even that *one only man*, who had been previously instructed to look for a certain almost microscopic mark on the floor."

APPENDIX.

I.—DERIVATION OF THE TERM PYRAMID. (*Page* 219.)

Professor Smyth suggests the origin of the term Pyramid from the two Coptic words, "*pyr*," "division," and "*met*," "ten." This derivation, which he first heard of in Cairo, is, he believes, a significant appellation for a metrological monument such as the Great Pyramid, and coincides with its five-sided, five-cornered, etc., features (see anteriorly, p. 255) and decimal divisions. But surely a name, which in this metrological and arithmetical view of "powers and times of ten and five," meant *division into ten*, and which divisional metrological ideas applied, according to Professor Smyth, to one pyramid only, namely the Great Pyramid of Gizeh, was not likely to have been applied as a general term to all the other pyramidal structures in Egypt—not one of which had, according to Professor Smyth himself, anything whatsoever of this metrological or divisional character in their composition and object. It is not likely that all these structures should have been named from a series of qualities supposed to belong to *one;* but altogether hidden and concealed, in these early times, even in that one pyramid, being for the information of future times and generations.

In a similar spirit of exclusiveness, Mr. John Taylor derives the word pyramid from the two Greek words πυρος, *wheat*, and μετρον, *measure*—apparently in the belief that the coffer or sarcophagus within one pyramid (the Great Pyramid) was intended as a chaldron measure of wheat—though none of the sarcophagi, in any of the many other royal pyramidal sepulchres of Egypt, were at all intended for such standard measures; and although, according to Mr. Taylor's theory, the Greeks, too, who out of their own language applied the term of Pyramid, or Wheat-Measurer, to all these structures,—never dreamed of the Great Pyramid or of any other of them having locked up in one of its concealed chambers a supposed standard measure of capacity of wheat, water, etc., for all nations and all times.

Fifteen centuries ago, Ammianus Marcellinus derived the word pyramid from another Greek word πυρ, *fire;* because, as he argues, the Egyptian Pyramid rises to a sharp pointed top, like to the form of a fire or flame. This derivation, which, of course, excludes the mathematical idea of the sides of the pyramid being a series of flattened triangles that meet in a point at the apex, has been adopted by various authors.

Keats, the poor surgeon, but rich poet, who died at Rome at the early age of twenty-six, was buried in the beautiful Protestant Cemetery there, amid the ruins of the Aurelian Walls. His grave is surmounted by a pyramidal tomb, which Petrarch romantically ascribed to Remus, but which antiquarians generally accord, in conformity with the inscription which it bears, to Caius Cestius, a tribune of the people, who is remembered for nothing else than his sepulchre. In his elegy of Adonais, Shelley, in alluding to the resting-place of Keats beside this remarkable monument, brings in, with rare poetical power, the idea of the word pyramid being derived from πυρ, and signifying the shape of flame :—

And one keen *pyramid* with edge sublime,
Pavilioning the dust of him who planned
This refuge for his memory, doth stand
Life *flame transformed to marble.*[1]

If the word pyramid is of Greek origin, the suggestion of that able writer and scholar, Mr. Kenrick of York, is probably more true, viz. that the term πυραμις (from πυρος, wheat, and μελιτος, honey) was applied by

[1] Shelley himself is now interred in the same cemetery, near the pyramid of Cestius, and a little above the grave of Keats.

the Greeks to a pointed or cone-shaped cake, used by them at the feasts of Bacchus (as shown on the table at the reception of Bacchus by Icarus; see Hope's *Costumes*, vol. ii. p. 224), and when they became acquainted with the Pyramids of Egypt, they, in this as in other instances, applied a term to a thing till then unknown, from a thing well known to them; in the very same way as they applied to the tall pointed monoliths peculiar to Egypt, the word obelisk—no doubt a direct derivation from the familar Greek word ὀβελος, a *spit*.

For a learned discussion on various other supposed origins of the word pyramid, see Jomard, in the *Description de l'Egypte*, vol. ii. p. 213, etc.

II.—ARCHAIC CIRCLE AND RING SCULPTURES. (*Page* 222.)

Representations of incised cups, rings, circles, and spirals, are found on stones connected with other forms of ancient sculpture besides chambered barrows or cairns,—as on the lids of stone cists, megalithic circles, etc.; and, from this connection with the burial of the dead, these antique sculpturings were possibly of a religious character. In a work on "Archaic Sculpturings of Cups, Rings, etc. upon Stones and Rocks of Scotland, England, and other Countries," published last year by the author of the present communication, it was further argued that they were probably also ornamental in their character, in a chapter beginning as follows:—

"Without attempting to solve the mystery connected with these archaic lapidary cups and ring cuttings, I would venture to remark that there is one use for which some of these olden stone carvings were in all probability devoted—namely, ornamentation. From the very earliest historic periods in the architecture of Egypt, Assyria, Greece, etc., down to our own day, circles, single or double, and spirals, have formed, under various modifications, perhaps the most common fundamental types of lapidary decoration. In prehistoric times the same taste for circular sculpturings, however rough and rude, seems to have swayed the mind of archaic man. This observation as to the probable ornamental origin of our cup and ring carvings holds, in my opinion, far more strongly in respect to some antique stone cuttings in Ireland and in Brittany, than to the ruder and simpler forms that I have described as existing in Scotland and England. For instance, the cut single and double volutes, the complete and half-concentric circles, the zig-zag, and other patterns which cover almost entirely and completely some stones in those magnificant though rude western Pyramids that constitute the grand old mausolea of Ireland and Brittany, appear to be, in great part at least, of an ornamental character, whatever else their import may be."

In a communication on the Great Pyramid, made to the Royal Society 16th December 1867, Professor Smyth most unexpectedly, and quite out of his way, took occasion to criticise severely the remarks contained in the preceding extract, on two grounds :

First, He laid down that the term pyramid was misapplied, as the term referred only to figures and structures of a special mathematical form ; being apparently quite unaware that, as shown in the text and notes, pp. 219 and 220, it was often applied archæologically to sepulchral mounds and erections that were not faced, and which did not consist of a series of triangles meeting in an apex.

Secondly, He objected to the statement that, "from the very earliest historic periods in the architecture of Egypt, Assyria, Greece, etc., circles and spirals, or modifications of them, constituted perhaps the most common fundamental types of lapidary decoration ;" because, though circles, spirals, etc., occurred in the later architecture of Thebes, etc., yet in the Great Pyramid of Gizeh no such decorations were to be found, nor, indeed, lapidary decorations of any other kind. Cheops, the builder of the Great Pyramid, was, according to Manetho, "arrogant towards the gods." Was it this spirit of religious infidelity or scepticism that led to the rejection of any ornamentation? Professor Smyth notices what he himself terms an "ornament," "a most unique thing certainly," on the upper stone of what Greaves calls "the granite leaf" portcullis, in the interior of the Great Pyramid (ii. 100), and he represents it, it is now said erroneously in plate xii. as a portion of a double circle instead of a general raised elevation.[1]

All the other Pyramids of Gizeh seem, like the Great Pyramid, wonderfully free from lapidary decorations on their interior walls, the exteriors of all of them being now too much dilapidated to offer any distinct proof in relation to the subject; though in Herodotus' time there were hieroglyphics, at least on the external surface of the Great Pyramid. The whole surface of the basalt sarcophagus in the Third Pyramid, or that of Mycerinus, was sculptured. "It was," to use the words of Baron Bunsen, "very beautifully carved in compartments, in the Doric style" (vol. ii. 168). This carving, in the well-known carpentry form, was, according to Mr. Fergusson, a representation of a palace (*Hand-book of Architecture*, p. 222).

Fragments, however, of lapidary sculpture have been found among the ruins of Egyptian pyramids supposed to be older than those of Gizeh, or than their builders, the Memphite kings of the *fourth* dynasty. Thus one of the most able and learned of modern Egyptologists, Baron Bunsen, has written at some length to show that the great northern brick pyramid of Dashoor belongs to the preceding or *third* dynasty of kings. Colonel Vyse and Mr. Perring, when digging among its ruins, discovered two or three fragments of sculptured casing and other stones, with a few pieces presenting broken hieroglyphic inscriptions. One of the ornamented fragments represents a row of floreated-like decorations, and each decoration shows on its side a concentric circle, consisting of three rings,—the

[1] In vol. i. p. 365, this "raised ornament" is described as "a very curious, and, for the Pyramid, perfectly unique adornment, of a semicircular form, raised about one inch above the general surface, and bevelled off on either side and above," etc.

whole ornament being one which is found in later Egyptian eras, not unfrequently along the tops of walls in the interior of chambers, etc. Mr. Perring represents this fragment of sculpturing from the brick Pyramid of Dashoor, in his folio work, *The Pyramids of Gizeh*, plate xiii. Fig. 7. Hence among the very earliest Egyptian lapidary decorations we have, as in other countries, the appearance of the simple circular ornamentation.

Besides, more complex circular and spiral decorations, in the form of the well-known guilloche and scroll, were made use of in Egypt during the sixth dynasty, or immediately after the Memphite dynasty that reared the larger Pyramids of Gizeh. Thus, speaking of the ancient Egyptian architectural decorations, Sir J. Gardner Wilkinson observes—"The Egyptians did not always confine themselves to the mere imitation of natural objects for ornament; and their ceilings and cornices offer numerous graceful fancy devices, among which are the guilloche, miscalled Tuscan borders, the chevron, and the scroll patterns. They are to be met with in a tomb of the time of the sixth dynasty; they are therefore known in Egypt many ages before they were adopted by the Greeks, and the most complicated form of the guilloche covered a whole Egyptian ceiling, upwards of a thousand years before it was represented on those comparatively late objects found at Nineveh."—*Popular account of the Ancient Egyptians* ii. 290.

III.—ERA OF THE ARABIAN HISTORIAN, IBN ABD AL HAKM. (*Page* 236.)

Professor Smyth owns that the grooves and pin holes which the coffer in the King's Chamber presents, were (to use his own words) "in fact to admit a sliding sarcophagus cover or lid" (see *ante*, p. 236, footnote). But in his recent communication to the Royal Society on the 20th April, he doubted Al Hakm's account of the mummy having been actually found in the sarcophagus when the King's Chamber was first entered by the

Caliph Al Mamoon, in the ninth century, arguing, on the authority of a Glasgow gentleman, that the historian himself, Al Hakm, did not live for three or four centuries afterwards, and, therefore, could not be relied upon. But all this reasoning or assertion is simply a mistake. In a late letter (7th April), Dr. Rieu of the British Museum,—the chief living authority among us on any such Arabic question,—writes, "The statement relating to Al Mamoon's discovery could hardly rest on a better authority than that of Ibn Abd Al Hakm; for not only was he a contemporary writer (having died at Old Cairo, A.H. 269, that is, thirty-eight years after Al Mamoon's death), but he is constantly quoted by later writers as an historian of the highest authority. You will find a notice of him in Khallikan's *Biographical Dictionary*, vol. ii. etc." He was a native of Egypt, and chief of the Shafite sect. Born in A.D. 799, he died in A.D. 882, or at the age of 83.

IV.—LENGTH OF THE SARCOPHAGUS IN THE KING'S CHAMBER. (*Page* 236.)

M. Jomard, in the *Description de l'Egypte*, drawn up by the French Academicians, remarks in vol. ii. p. 182, that looking to the length of the cavity or interior of the sarcophagus in the King's Chamber, that it could not hold within it a cartonage or mummy case, enclosing a man of the ordinary height. This statement proceeds entirely upon a miscalculation. The length of the interior or cavity of the sarcophagus is six and a half English feet; and the average stature of the ancient Egyptians, "judging from their mummies, did not" observes Mr. Kenrick, "exceed five feet and a half." (See his *Ancient Egypt*, vol. i. p. 97.) The space thus left, of one foot, is much more than sufficient for the thickness of the two ends of a cartonage or mummy case; and the embalmed body was generally, or indeed always, closely packed within them. The length of the coffin was, long ago, quaintly observed Professor Greaves, "large enough to contain a most potent and dreadful monarch being dead, to whom, living, all Egypt was too strait and narrow a circuit" (*Works*, i. p. 131).

V.—MEMORANDUM ON THE CUBIT OF MEMPHIS AND THE SACRED CUBIT, BY SIR HENRY JAMES. (*Page* 242.)

Sir Isaac Newton says, "for the precise determination of the cubit of Memphis I should choose to pitch upon the length of the chamber in the middle of the Pyramid, where the king's monument stood, which length contained 20 cubits, and was very carefully measured by Mr. Greaves." (*See* vol. ii. p. 362 of Professor Smyth's *Life at the Pyramids*, etc.)

Greaves' measures of the King's chamber are given at p. 335, vol. ii. of the same work.

The length of the chamber on the south side, he says, is

34·380 feet = 20 cubits.
17·190 „ = 10 cubits.
12

206·280 inches = 10 cubits.
and 20·628 „ = 1 cubit of Memphis;

and Newton himself says, at p. 360, vol. ii. *Life at the Pyramids*,—

"The cubit of Memphis of 1·719 English feet,"
12

or 20·628 inches,

and, therefore, there can be no possible doubt but that this is Newton's determination of the length of the cubit of Memphis.

But Newton goes on to say in the same page, the cubit "double the length of $12\frac{3}{8}$ English inches (= 24·75 inches) will be to the cubit of Memphis as 6 to 5."

Therefore, if we add $\frac{1}{5}$ to 20·628 inches.
4·126

we have 24·754

as Newton's determination of the length of the Sacred Cubit.

Newton's determinations are therefore—

Length of Sacred Cubit 24·754 inches.
„ Cubit of Memphis 20·628 „

The cubit measured by Mersennus (*see* p. 362, vol. ii. *Life at the Pyramids*) was 23¼ Paris inches, and Mr. Greaves estimated the Paris foot as equal to 1·068 of the English foot; therefore 23·25 + 1·068 = 24·831 was the length of this cubit, if we take Greaves' proportion of the Paris to the English foot; but by the more exact determination of the proportion of the Paris to the English foot made at the Ordnance Survey Office, Southampton, it is found to be as 1 to 1·06576 and 23·25 + 1·06576 = 24·780 English inches, which differs only in excess ·026 from the length of the Sacred Cubit determined by Newton.

The double Royal Cubit of Karnak, which is in the British Museum, was found by Sir Henry James to measure 41·398 inches; the length of the single cubit was therefore 20·699 inches, and differs only in excess ·071 inches from the length of the cubit of Memphis, as determined by Newton.

It will be observed that the lengths of the cubits derived by Newton from the length of the King's chamber are shorter than the measured lengths of the cubits which have come down to us. But if

we add $\frac{1}{5}$ or	= 4·140	to the length of the
Karnak cubit	= 20·699,	
we have	24·839	for the Sacred Cubit.
The one measured by Mersennus	= 24·780	and the
mean of the two	= 24·810,	whilst the
length derived by Newton was	= 24·754,	showing
a difference of only	·056	between the

length of the Sacred Cubit derived from the actual lengths of the two cubits which have come down to us, and the length of the Sacred Cubit derived by Newton from the length of the King's chamber.

The method adopted by Professor P. Smyth, to find the length of the Sacred Cubit, in p. 458, vol. ii. *Life at the Pyramids*, is also wrong in principle. He has no right to take the means between the limits of approach, or to say that the Sacred Cubit was, according to Sir Isaac Newton, 25·07 inches, when, as I have shown in his own words, Sir Isaac says it was 24·754 inches.

VI.—PROFESSOR SMYTH'S RECENT COMMUNICATION TO THE ROYAL SOCIETY ON 20TH APRIL 1868.

It has been already stated (see footnote, p. 248) that, on the 20th April Professor Smyth brought before the Royal Society a new communication on the pyramids, the principal part of which consisted of a criticism upon the preceding observations, and a defence of his hypotheses regarding the Great Pyramid. His chief criticisms related to points already adverted to, and answered in footnotes, pp. 234, 248, etc. In addition, he expressed great dissatisfaction that the quotation from Sprenger, in Vyse's Work, quoted in footnote, p. 237, was not extended beyond the semicolon in the original, at which the quotation ends, and made to embrace the other or latter half of the sentence, viz., ". . . . ; and that they appear to have repeated the traditions of the ancient Egyptians, mixed up with fabulous stories and incidents, certainly not of Mahometan invention."[1] But this

[1] The whole sentence runs thus, and is punctuated thus :—"It may be remarked that the Arabian authors have given the same accounts of the pyramids with little or no variation, for above a thousand years ; and that they appear to have repeated the traditions of the ancient Egyptians, mixed up with fabulous stories and incidents, certainly not of Mahometan invention." Vol. iii. p. 328.

latter half, or the traditions about the pyramid builders, Surid, Ben Shaluk, Ben Sermuni, etc., who lived "before the Flood," etc. etc., did assuredly not require to be quoted, as they had really nothing whatever to do with the object under discussion—viz., the opening of the sarcophagus under the Caliph Al Mamoon, and the accounts or history of the pyramids, as given by Arabian authors themselves.

In the course of this communication to the Royal Society, Professor Smyth did not allude to or rescind the erroneous table and calculations from Sir Isaac Newton regarding the Sacred Cubit, printed and commented upon in some of the preceding pages (see *ante*, p. 244, etc.) But, at the end of the subsequent discussion he handed round, as a printed "Appendix" to his three volume work, a total withdrawal of this table, etc., and in this way so far confessed the justice of the exposition of his errors on this all-vital and testing point in his theory of the Sacred Cubit, as given in p. 243, etc., of the present essay. He attributes his errors to "an unfortunate misprinting of the calculated numbers;" and (though he does not at all specialise what numbers were thus misprinted) he gives from Sir Isaac Newton's Dissertation on the Sacred Cubit a new and more lengthened table instead of the old and erroneous table. For this purpose, instead of selecting as he did, without any attempted explanation in his old table, *only five* of Sir Isaac Newton's estimations or "methods of approach," he now, in his new table, takes *seven* of them to strike out new "means." The simple "mean" of all the seven quantities tabulated —as calculated, in the way followed, in his first published table—is 25·47 British inches; and the "mean" of all the seven means in the Table is 25·49 British inches. Unfortunately for Professor Smyth's theory of the Sacred Cubit being 25·025 British inches, either of these numbers makes the Sacred Cubit nearly half a British inch longer than his avowed standard of length—an overwhelming difference in any question relating to a *standard* measure. What would any engineer, or simple worker in metal, wood, or stone, think of an alleged *standard* measure or cubit which varied so enormously from its own alleged length? But, surely, such facts and such results require no serious comment.

In this, his latest communication on the Pyramids, Professor Smyth also offered some new calculations regarding the measurements of the interior of the broken stone coffin standing in the King's Chamber. Formerly (1864), he elected the cubic capacity of this sarcophagus to be 70,900 "pyramidal" cubic inches; latterly he has elected it to be 71,250 cubic inches. According, however, to his own calculations, he found, practically, that it measured neither of these two numbers; but instead of them 71,317 pyramidal inches (*see* vol. iii. p. 154). The capacity of the interior of this coffin does not hence correspond at all to the supposititious standard of 71,250 pyramidal cubic inches; but in order to make it appear to do so he has now struck a "mean" between the measurement of the interior of the vessel and some of the measurements of its exterior, in a way that was not easily comprehensible in his demonstration. But what other hollow vessel in the world, and with unequal walls too (*see* p. 233), had the capacity of its interior ever before attempted to be altered and rectified by any measurements of the size of its exterior? What, for example, would be thought of the very strange proposition of ascertaining and determining the capacity of the interior of a pint, a gallon, a bushel, or any other such standard measure by measuring, not the capacity of the interior of the vessel, but by taking some kind of mean between that interior capacity and the size or sizes of the exterior of the vessel? According to Messrs. Taylor and Smyth, this standard measure—along with other supposed perfect metrological standards—in the Great Pyramid is "of an origin higher than human," or "divinely inspired;" and yet it has proved so incapable of being readily measured, and hence used as a standard, that hitherto it has been found impossible to make the *actual* capacity of this coffer to correspond to its standard theoretical or supposititious capacity; whilst even its standard theoretical capacity has been declared different by different observers, and even at different times by the same observer, as shown previously at p. 231.

VII.—METROLOGICAL TABLES AND TESTS OF THE EUROPEAN RACES. (*See* p. 238.)

Professor Smyth believes that among the nations of Europe the metrology used will be found closer and closer to the Hebrew and "Pyramid" standards, according to the amount of Ephraimitic blood in each nation. He further inclines to hold, with Mr. Wilson, that the Anglo-Saxons have no small share of this Israelitish blood, as shown in their language, and in their weights and measures, etc. After giving various Tables of the metrological standards of different European nations, Professor Smyth adds, "It is not a little striking to see all the Protestant countries standing first and closest to the Great Pyramid; then Russia, and her Greek, but freely Bible-reading church; then the Roman Catholic lands; then, after a long interval, and last but one on the list, France with its metrical system—voluntarily adopted, under an atheistical form of government, in place of an hereditary pound and ancient inch, which were not very far from those of the Great Pyramid; and last of all Mahommedan Turkey." Subsequently, when speaking of British standards of length, etc., Professor Smyth remarks,—"But let the island kingdom look well that it does not fall; for not only has the 25·344 inch length not yet travelled beyond the region of the Ordnance maps,—but the Government has been recently much urged by, and has partly yielded to, a few ill-advised but active men, who want these invaluable hereditary measures (preserved almost miraculously to this nation from primeval times, for apparently a Divine purpose) to be instantly abolished *in toto*,—and the recently atheistically-conceived measures of France to be adopted in their stead. In which case England would have to descend from her present noble pre-eminence in the metrological scale of nations, and occupy a place almost the very last in the list; or next to Turkey, and in company with some petty princedoms following France, and blessed with little history and less nationality.

'How art thou fallen from heaven, O Lucifer, son of the morning!' might be then, indeed, addressed to England with melancholy truth. Or more plainly (Professor Smyth adds), and in words seemingly almost intended for such a case, and uttered with depressing grief of heart, 'O Israel, thou hast destroyed thyself!'" (Professor Smyth's *Life and Work at the Great Pyramid*, 1867, vol. iii. p. 598.)

In his previous work in 1864, Professor Smyth denounced also, in equally strong terms, the French decimal system of metrology, considering it as—to use his own words—"precisely one of the most hearty aids which Satan, and traitors to their country, ever had presented to their hands." (*Our Inheritance in the Great Pyramid*, p. 185, etc.)

END OF VOL. I.

Printed by R. & R. CLARK, *Edinburgh*.

www.ingramcontent.com/pod-product-compliance
Lightning Source LLC
Chambersburg PA
CBHW021946220326
41599CB00012BA/1193

* 9 7 8 3 7 4 3 3 5 8 4 2 3 *